NONDESTRUCTIV.
CHARACTERIZATION
OF COMPOSITE MEDIA

HOW TO ORDER THIS BOOK

BY PHONE: 800-233-9936 or 717-291-5609, 8AM-5PM Eastern Time

BY FAX: 717-295-4538

BY MAIL: Order Department
Technomic Publishing Company, Inc.
851 New Holland Avenue, Box 3535
Lancaster, PA 17604, U.S.A.

BY CREDIT CARD: American Express, VISA, MasterCard

NONDESTRUCTIVE CHARACTERIZATION OF COMPOSITE MEDIA

Ronald A. Kline

Professor
School of Aerospace and Mechanical Engineering
The University of Oklahoma
Norman, Oklahoma, U.S.A.

TECHNOMIC
PUBLISHING CO., INC.
LANCASTER · BASEL

Nondestructive Characterization of Composite Media
a **TECHNOMIC** publication

Published in the Western Hemisphere by
Technomic Publishing Company, Inc.
851 New Holland Avenue
Box 3535
Lancaster, Pennsylvania 17604 U.S.A.

Distributed in the Rest of the World by
Technomic Publishing AG

Printed in the United States of America
10 9 8 7 6 5 4 3 2 1

Main entry under title:
    Nondestructive Characterization of Composite Media

A Technomic Publishing Company book
Bibliography: p. 177
Includes index p. 185

Library of Congress Card No. 92-61121
ISBN No. 0-87762-925-0

To my wife, Lynnell, with heartfelt thanks, for her unwavering support throughout this project.

ACKNOWLEDGEMENT

The author is greatly indebted to several of his former and present graduate students for their assistance in many phases of this work: R. Adams, M. Doroudian, C. Hsiao, R. Kulathu, L. Moore, S. Sahay, Y. Wang and especially C. Sullivan for her careful proofreading of the final manuscript. Also, the patience and care of Ms. Rose Benda in typing the manuscript are gratefully acknowledged.

The major objective of this monograph is to present the basic concepts of wave propagation in anisotropic media, and to show how these concepts can be applied to the quantitative, nondestructive evaluation of composite media. The emphasis here is on quantitative rather than qualitative NDE. In a typical composite test, one generally looks for anomalies in the echo pattern to indicate the presence of a defect in the sample. While this may be a useful test to find gross, large-scale defects in the composite, it provides little fundamental information as to the nature of the defect, its effect on mechanical properties of the material, or the expected performance of the part in its service environment. As this type of test is qualitative in nature, accept/reject criteria for composites are not as objective as one might desire.

In this work, the subject of composite NDE is approached from a standpoint somewhat different from that of conventional part inspection based on an analysis of echo amplitudes. Here we seek to examine, in detail, those features of ultrasonic wave propagation that can yield important quantitative information about composite materials, their microstructures, and their mechanical properties. Of partial importance in this regard is the anisotropy of composite response. The presence of a preferentially aligned reinforcing medium embedded in the composite means that composite response is directionally dependent. The ability to launch sound waves in multiple directions in a composite gives a unique way of characterizing material anisotropy in a quantifiable and experimentally tractable fashion. This information is often overlooked in conventional NDE test procedures.

The monograph can be divided into two sections. The initial section is devoted to developing the theoretical basis for sound wave propagation in composite media. Particular attention is devoted to describing the features of wave propagation in anisotropic media which differ from those corresponding to isotropic media. Representation surfaces, such as the slowness surface, are introduced to provide a means of visualizing the effects of anisotropic response on material properties and wave propagation. Also, the unique aspects of energy propagation in anisotropic media are described in detail. Since bulk and guided waves have been used extensively for composite material characterization, the basic governing equations for both bulk and guided waves are presented.

The remainder of the monograph deals with practical applications. It is a review of experimental technique development for composites. As ultrasonic velocity is directly related to elastic response, the major thrust of this research has been the development of procedures for elastic property reconstruction from bulk and/or guided wave measurements. Of late there has been a tremendous amount of effort devoted to this problem and, as a result, several alternative procedures are available to achieve this goal. It should be mentioned that the ability to nondestructively characterize the mechanical properties in a quantitative way is not necessarily an end in itself. Rather, these measurements can be coupled with other material evaluation techniques such as FEM to enable the one to evaluate the response of the actual as-fabricated component; not the behavior of an idealized, homogeneous part. Also, through the use of composite micromechanics, these measurements (either after fabrication or on-line during processing) can be used for quantitative characterization of composite microstructure. As time progresses, it is expected that quantitative NDE techniques, such as those described in this monograph, will grow in importance to the composites industry.

R. A. Kline
Norman, OK

ρ = density
$\boldsymbol{u}$ = particle displacement (vector)
$\underset{\sim}{\sigma}$ = stress tensor (2nd order tensor)
$\underset{\sim}{\epsilon}$ = strain tensor (2nd order tensor)
$\underset{\approx}{C}$ = stiffness tensor (4th order tensor)
$\boldsymbol{l}$ = wave normal
α = polarization (vector)
k = wave number
ω = frequency
$\boldsymbol{m}$ = slowness (vector)
$\boldsymbol{k}$ = wave vector
ν = surface normal (vector)
E = total energy
w = kinetic energy density
ϕ = strain energy density
$V = V_{\text{phase}}$ = phase velocity
$\boldsymbol{F}$ = energy flux density (vector)
$\boldsymbol{S}$ = velocity of energy flow
$V_{\text{group}} = |\boldsymbol{S}|$ = group velocity
λ, μ = Lamé constants
$\underset{\sim}{C}$ = stiffnesses (in reduced notation, 6×6 matrix)
$\underset{\sim}{L}$ = operator representation for equations of motion
$\underset{\sim}{\delta}$ = identity matrix
V_f = fiber volume fraction

$$\cdot = \frac{\partial}{\partial t}, \qquad \cdot\cdot = \frac{\partial^2}{\partial t^2}, \text{ etc.} \qquad ' = \frac{\partial}{\partial y}, \qquad '' = \frac{\partial^2}{\partial y^2}, \text{ etc.}$$

For simplicity, indicial notation is used wherever possible. This means that quantities are usually represented in terms of their components, i.e.,

V_i represents the ith component of the vector V,

Q_{ij} represents the ijth component of the second order tensor Q, etc.

Also, Einsteinian summation convention is followed in this text. This means that whenever repeated indices are observed that a summation of that index from 1 to 3 is understood, e.g., the dot product of two vectors a and b can be represented as

$$a \cdot b = a_i b_i \Longrightarrow \sum_{i=1}^{1} a_i b_i$$

CHAPTER 1

Basic Governing Equations

The basic governing equations for elastic wave propagation in an anisotropic medium can be derived directly from the equations of motion for a continuum (in the absence of body forces)

$$\rho \ddot{u}_i = \sigma_{ij,j} \tag{1.1}$$

Assuming a linearly elastic constitutive equation for the material

$$\sigma_{ij} = C_{ijkl}\epsilon_{kl} \tag{1.2}$$

we have

$$\rho \ddot{u}_i = C_{ijkl}\epsilon_{kl,j} \tag{1.3}$$

Substituting the strain-displacement relationship (for small displacements)

$$\epsilon_{kl} = \frac{1}{2}(u_{k,l} + u_{l,k}) \tag{1.4}$$

into Equation (1.3), we have

$$\rho \ddot{u}_i = \frac{1}{2}[C_{ijkl}u_{k,lj} + C_{ijkl}u_{l,kj}] \tag{1.5}$$

However, the symmetry of the strain and stress tensors requires

$$C_{ijkl} = C_{ijlk} = C_{jilk} \tag{1.6}$$

Working with the second term in Equation (1.5), we may interchange l and k

$$C_{ijkl} u_{l,kj} = C_{ijlk} u_{k,lj} \tag{1.7}$$

as (from the summation convention) l and k are dummy indices. Also, from the symmetry of the strain tensor

$$C_{ijlk} = C_{ijkl} \tag{1.8}$$

Therefore, we have

$$C_{ijkl} u_{l,kj} = C_{ijkl} u_{k,lj}$$

and Equation (1.5) assumes the simplified form

$$\rho \ddot{u}_i = C_{ijkl} u_{k,lj} \tag{1.9}$$

For harmonic plane wave propagation in a given direction in the material, we may assume the displacements to be given by the following expression

$$u_i = A_o \alpha_i e^{i[k(l_i x_i) - \omega t]} \tag{1.10}$$

where

A_o = displacement amplitude
α = polarization vector
k = wave number
l = direction of propagation
ω = frequency
$\dfrac{\omega}{k}$ = phase velocity

Substituting this representation for the displacement field into Equation (1.9) yields an eigenvalue problem where the eigenvalues yield the possible velocities of propagation in a given direction and the eigenvectors are the corresponding polarizations, i.e.,

$$(C_{ijkl} l_j l_l - \rho V^2 \delta_{ik}) \alpha_k = 0 \tag{1.11}$$

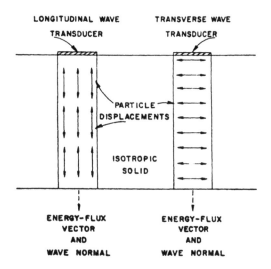

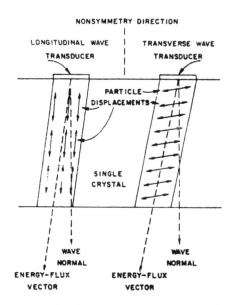

FIGURE 1.1 Schematic illustrating the differences in wave propagation in isotropic media (or anisotropic media in certain symmetry directions) and anisotropic media in nonsymmetry directions (after Green [1]).

also known as the Christoffel equation. Hence, in any given direction, three waves may be propagated. Waves can be distinguished by the relationship between the polarization vector and the direction of propagation. If

$$\alpha \cdot l = 1 \qquad (1.12)$$

the wave will be a pure mode longitudinal wave with particle displacements parallel to the wave normal. If

$$\alpha \cdot l = 0 \qquad (1.13)$$

the wave will be a pure mode transverse wave with particle displacements perpendicular to the wave normal. These waves are illustrated in Figure 1.1 [1]. For an arbitrary direction in an anisotropic medium, neither Equation (1.12) or (1.13) usually holds. These waves are usually classified as quasilongitudinal or quasitransverse depending on the magnitude of the component of the particle displacement vector in the direction of the wave normal.

CHAPTER 2

Wave Surfaces

Prior to the advent of digital computers, the algebraic complexity of wave propagation problems necessitated lengthy and time-consuming hand calculations for accurate solutions. Accordingly, geometric means were developed for approximate calculations. The variation in acoustic properties with direction in anisotropic media can be represented geometrically in a variety of ways. These representations remain useful in illustrating various aspects of wave propagation in anisotropic media.

Kriz and Ledbetter [2] have extensively studied the use of representation surfaces to describe the mechanical behavior of fiber-reinforced structures. The directional dependence of elastic properties can be easily visualized with such techniques. The variation of Young's modulus with direction in a unidirectionally reinforced composite is shown in Figure 2.1, as a function of fiber volume fraction (V_f). Here, the x_3 axis represents the direction of fiber reinforcement and θ represents the angle between the tensile axis and fiber reinforcement direction. One can calculate Young's modulus for any given θ and V_f from composite micromechanics. To construct the curves shown in the figure, one begins with a particular value of θ and calculates the value of Young's modulus in this direction. This defines a vector whose magnitude is given by Young's modulus and whose direction is determined from θ. The loci of the endpoints of all such vectors form a two-dimensional section of the elastic representation surface. By considering all possible directions, the entire three-dimensional surface can be developed in an analogous fashion. Note that the section for a zero fiber volume fraction is circular as the resin itself is isotropic. However, as additional fibers are added to the material, the stiffness increases, particularly in the fiber reinforcement direction. Similar plots may be developed for other important

5

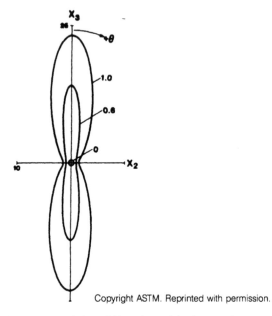

FIGURE 2.1 Representation of directional variation of Young's modulus in a unidirectional composite (after Kriz and Ledbetter [2]).

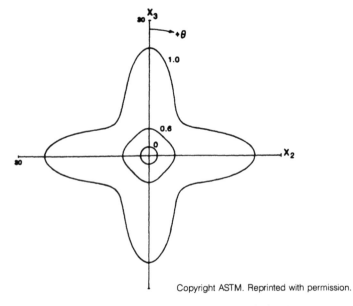

FIGURE 2.2 Representation of directional variation of shear modulus (G_{23}) in a unidirectional composite (after Kriz and Ledbetter [2]).

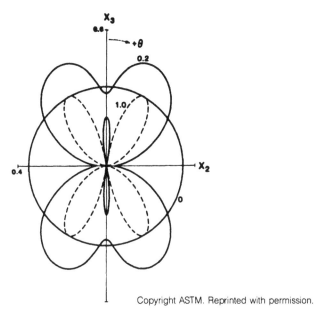

FIGURE 2.3 Representation of the directional variation of Poisson's ratio in a unidirectional composite (after Kriz and Ledbetter [2]).

material parameters such as the G_{23} shear modulus (Figure 2.2) and the ν_{23} Poisson's ratio (Figure 2.3).

For wave propagation, one can proceed in precisely the same fashion to develop curves that can be used to characterize the directional dependence of acoustic properties.

The most commonly used representation is the slowness surface, which has great utility in treating reflection-refraction phenomena in anisotropic media. We define the slowness vector in a given direction of propagation l as the reciprocal of the phase velocity, i.e.,

$$m = \frac{1}{\omega}\,k = \frac{1}{V}\,l$$

where

k = wave vector = $|k|l$
$k = |k|$ = wave number
l = wave normal

The slowness surface represents the locus of the endpoints of the slowness vector. For an isotropic medium, the slowness surface consists of three concentric spherical sheets. The inner sheet represents the longitudinal wave and the outer sheets are two coincident sheets associated with shear wave propagation. The sheets are spherical because the velocities in an isotropic medium are independent of the propagation direction. The two shear sheets coincide since the two velocities of shear wave propagation are identical. For anisotropic media, there are three distinct sheets of arbitrary shape. The shape of the slowness surface is an important factor in determining the nature of reflected and transmitted waves in anisotropic media, as will be shown in the following discussion.

Consider a plane longitudinal wave incident upon the boundary of a composite panel. Here, for clarity, we take the incident medium to be water, as for an immersion experiment. The incident wave may be represented as

$$u_{in} = A_{in}e^{i\omega^{in}(m_k^{in}x_k - t)} \tag{2.1}$$

Similarly, the reflected wave may be represented as

$$u_{re} = A_{re}e^{i\omega^{re}(m_k^{re}x_k - t)} \tag{2.2}$$

and for the transmitted waves, we have

$$u_i = A_i e^{i\omega_k^i(m_k^i x_k - t)} \tag{2.3}$$

where the superscript i is used to differentiate between the various possible transmitted waves. In order to satisfy the boundary conditions at the interface, the frequencies of all waves must be equal, i.e.,

$$\omega = \omega^{in} = \omega^{re} = \omega^i \tag{2.4}$$

and

$$m_k^{in} x_k |_{\text{boundary}} = m^{re} x_k |_{\text{boundary}} = m_k^i x_k |_{\text{boundary}} \tag{2.5}$$

Henneke [3] has shown that Equation (2.5) is equivalent to

$$a_i = \epsilon_{ijk} m_j^{in} v_k = \epsilon_{ijk} m_j^{re} v_k = \epsilon_{ijk} m_j^i v_k$$

where

v = normal to interface
a = constant vector quantity

This statement is precisely Snell's law. To see this, let

$$\nu = \begin{pmatrix} 1 \\ 0 \\ 0 \end{pmatrix}$$

and

$$\boldsymbol{m}^{in} = \frac{1}{V_w} \begin{pmatrix} \cos \theta^{in} \\ 0 \\ \sin \theta^{in} \end{pmatrix}$$

as is appropriate for this geometry so that

$$\boldsymbol{a} = \begin{pmatrix} 0 \\ \dfrac{\sin \theta^{in}}{V_w} \\ 0 \end{pmatrix}$$

Equation (2.5) then becomes

$$\begin{pmatrix} 0 \\ \dfrac{\sin \theta^{in}}{V_w} \\ 0 \end{pmatrix} = \begin{pmatrix} 0 \\ m_3^{re} \\ -m_2^{re} \end{pmatrix} = \begin{pmatrix} 0 \\ m_3^{i} \\ -m_2^{i} \end{pmatrix} \qquad (2.6)$$

This requires

$$m_2^{re} = m_2^{i} = 0 \qquad (2.7)$$

Then, since l must be a unit vector, we may use the following representation:

$$\boldsymbol{m}^{re} = \frac{1}{V_w} \begin{pmatrix} -\cos \theta_{re} \\ 0 \\ \sin \theta_{re} \end{pmatrix} \quad \text{and} \quad \boldsymbol{m}^{i} = \frac{1}{V_i} \begin{pmatrix} \cos \theta^{i} \\ 0 \\ \sin \theta^{i} \end{pmatrix} \qquad (2.8)$$

where the negative sign in the slowness vector of the reflected wave is included to indicate that it is propagating away from the interface.

Then, we have

$$\frac{\sin \theta^{in}}{V_w} = \frac{\sin \theta^{re}}{V_w} = \frac{\sin \theta^i}{V_i} \qquad (2.9)$$

which is, of course, Snell's law. Letting

$$\boldsymbol{b} = \boldsymbol{v} \times \boldsymbol{a} = \begin{pmatrix} 0 \\ 0 \\ \dfrac{\sin \theta^{in}}{V_w} \end{pmatrix} \qquad (2.10)$$

the reflection–refraction problem may be represented geometrically by Figure 2.4. Here, the slowness surface sections and slowness vectors for the various waves are presented. We seek solutions to the equations of motion such that the projection of the slowness vector on the boundary ($|\boldsymbol{b}|$) is the same for all waves under consideration. For an isotropic medium, this means that there will be two possible reflected or refracted waves. For an anisotropic medium, the number and nature of the generated waves depend upon the shape of the slowness surface. The usual case, three possible refracted waves, is shown in Figure 2.4. However,

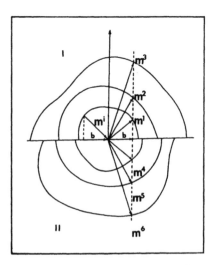

FIGURE 2.4   Reflection/refraction at an interface between two anisotropic materials (after Henneke). **m** = slowness vector.

as Henneke demonstrated, other configurations of the slowness surface are possible [3].

Other geometric representations can be developed from the governing equations. Instead of using the reciprocal of the velocity for our radius, we can use the phase velocity directly and obtain the velocity surface. Alternatively, group velocity can be used instead of phase velocity and slowness to obtain a surface conventionally known as the wave surfaces. This will be fully explored in the next section.

CHAPTER 3

Energy Propagation

The total energy in a given volume in a deformed elastic body is given by the sum of the kinetic and potential energies and may be expressed as

$$E = \int W d\tau + \int \phi d\tau = \int e d\tau \qquad (3.1)$$

where

$$W = \text{Kinetic Energy Density} = \frac{1}{2} \rho \dot{u}_i^2$$

$$\phi = \text{Potential (or Strain) Energy Density} = \frac{1}{2} \sigma_{ij} \epsilon_{ij} = \frac{1}{2} C_{ijkl} \epsilon_{ij} \epsilon_{kl}$$

For wave propagation, we are principally interested in the propagation of energy, i.e., the transport of energy across the wave front. Following Fedorov [4], this can be evaluated by considering the rate of change of the total energy (ΔE) in the volume τ, i.e.,

$$\frac{dE}{dt} = \int \frac{\partial W}{\partial t} d\tau + \int \frac{\partial \phi}{\partial t} d\tau \qquad (3.2)$$

But

$$\frac{\partial W}{\partial t} = \frac{\partial}{\partial t} \left(\frac{1}{2} \rho \dot{u}_i^2 \right) = \rho \dot{u}_i \ddot{u}_i \qquad (3.3)$$

13

and, via the chain rule

$$\frac{\partial \phi}{\partial t} = \frac{\partial \phi}{\partial \epsilon_{ij}} \frac{\partial \epsilon_{ij}}{\partial t} = \sigma_{ij} \frac{1}{2} (\dot{u}_{i,j} + \dot{u}_{j,i}) \tag{3.4}$$

Since the stress tensor must be symmetric and both i and j are dummy indices

$$\sigma_{ij} \dot{u}_{i,j} = \sigma_{j,i} \dot{u}_{j,j} = \sigma_{ij} \dot{u}_{j,i} \tag{3.5}$$

so we may write

$$\frac{\partial \phi}{\partial t} = \sigma_{ij} \dot{u}_{i,j} \tag{3.6}$$

Again, using the chain rule

$$(\sigma_{ij} \dot{u}_i)_{,j} = \sigma_{ij,j} \dot{u}_i + \sigma_{ij} \dot{u}_{i,j} \tag{3.7}$$

so

$$\frac{\partial \phi}{\partial t} = (\sigma_{ij} \dot{u}_i)_{,j} - \sigma_{ij,j} \dot{u}_i \tag{3.8}$$

Thus, we have

$$\frac{dE}{dt} = \int \rho \dot{u}_i \ddot{u}_i d\tau + \int (\sigma_{ij} \dot{u}_i)_{,j} d\tau - \int \sigma_{ij,j} \dot{u}_i d\tau \tag{3.9}$$

$$= \int (\rho \ddot{u}_i - \sigma_{ij,j}) \dot{u}_i d\tau + \int (\sigma_{ij} \dot{u}_i)_{,j} d\tau \tag{3.10}$$

where the first integral must vanish due to the equations of motion for a continuum. Using the divergence theorem, the second integral can be converted to a surface integral to yield

$$\frac{dE}{dt} = \int \sigma_{ij} \dot{u}_i n_j dS \tag{3.11}$$

This equation represents the energy crossing the wavefront/unit time. We define the energy flux density vector in the following fashion

$$\frac{dE}{dt} + \int F_j n_j dS = 0 \tag{3.12}$$

$$F_j = -\sigma_{ij} \dot{u}_i \tag{3.13}$$

which can be interpreted as the change in energy in a given volume that occurs via the flow or flux of energy (F) across the boundary surface for the volume or, for our purposes, the wave front. It is clear that energy does not necessarily propagate normal to the wave front; nor does energy generally propagate with the phase velocity V as defined earlier.

For plane wave propagation, we can develop the appropriate relationship for the velocity of energy propagation or group velocity by reexamining the energy for a wave field given by

$$u_i = A\alpha_i \exp\left[i\{k(l_i x_i) - \omega t\}\right]$$

$$= A\alpha_i \exp\left[i\psi\right] \tag{3.14}$$

Then, we may evaluate the kinetic energy densities as

$$W = \frac{1}{2}\rho \dot{u}_i \dot{u}_i$$

$$= -\frac{1}{2}\rho\omega^2 A^2 \exp\left[i2\psi\right] \tag{3.15}$$

and potential energy density as

$$\phi = \frac{1}{2}\sigma_{ij}\epsilon_{ij} = \frac{1}{2}C_{ijkl}u_{k,l}u_{i,j}$$

$$= -\frac{1}{2}C_{ijkl}A^2 k^2 \alpha_k \alpha_i l_l l_j \exp\left[i2\psi\right] \tag{3.16}$$

Note: The negative sign here is a consequence of the complex valued wavefield, but we can show from the equation of motion ($\sigma_{ij,j} = \rho \ddot{u}_i$) that

$$C_{ijkl} A k^2 \alpha_k l_l l_j \exp [i\psi] = \rho \omega^2 A \alpha_i \exp [i\psi] \qquad (3.17)$$

Multiplying both sides by $-u_i/2$, and using (3.14) we obtain

$$W = \phi$$

for plane wave propagation. Therefore we may represent the total energy as

$$E = 2 \int W d\tau = 2 \int \phi d\tau$$

$$= - \int \rho \omega^2 A^2 \exp [i2\psi] d\tau \qquad (3.18)$$

Similarly, the energy flux density vector will be given by

$$F_i = -\sigma_{ij} \dot{u}_j = C_{ijkl} A^2 \omega k l_l \alpha_k \alpha_j \exp [i2\psi] \qquad (3.19)$$

By taking the time average (one cycle) of the energy flux density and dividing it by the time average of the energy density, we obtain a vector expression for the velocity (S) of energy flow for wave propagation, i.e.,

$$S_i = \frac{\int_0^{2\pi/\omega} F_i dt}{\int_0^{2\pi/\omega} E dt} = \frac{C_{ijkl} A^2 \omega k l_l \alpha_j \alpha_k}{\rho A^2 \omega^2} = \frac{C_{ijkl} l_l \alpha_j \alpha_k}{\rho V} \qquad (3.20)$$

Of paramount importance is the fact that energy will propagate at the group velocity $V_{\text{group}} = |S|$ in the direction of S which may not necessarily coincide with the wave normal l as shown in Figure 3.1 [5]. This phenomena is known as energy flux deviation or beam skew. Alternatively, this result may also be obtained directly by realizing that the group velocity $|S|$ may be obtained from

$$\frac{\partial \omega}{\partial k} = S \quad \text{or} \quad \frac{\partial \omega}{\partial k_q} = S_q \quad \text{where} \quad k_q = k l_q \quad (3.21)$$

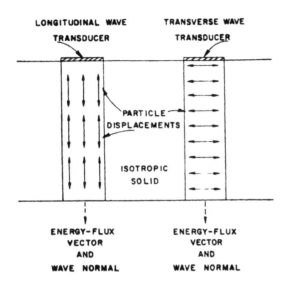

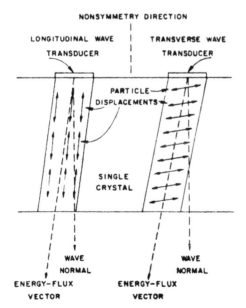

FIGURE 3.1 Illustration of energy flux deviation in anisotropic media (after Green [5]).

At this point it is convenient to work with ω^2 rather than ω. Using Equation (3.17), and the fact that α is a unit vector, we may write:

$$\omega^2 = \frac{C_{ijkl}}{\rho} k_l k_j \alpha_k \alpha_i \qquad (3.22)$$

Upon differentiation with respect to k_q

$$\frac{\partial \omega^2}{\partial k_q} = \frac{C_{ijkl}}{\rho} k_l \delta_{jq} \alpha_k \alpha_i + \frac{C_{ijkl}}{\rho} \delta_{lq} k_j \alpha_k \alpha_i$$

$$= \frac{C_{iqkl}}{\rho} k_l \alpha_k \alpha_i + \frac{C_{ijkq}}{\rho} k_j \alpha_k \alpha_i \qquad (3.23)$$

$$= 2\omega \frac{\partial \omega}{\partial k_q}$$

Using the symmetry of the stiffness tensor and the presence of dummy indices in the first expression

$$\frac{C_{iqkl}}{\rho} k_l \alpha_k \alpha_i = \frac{C_{kqij}}{\rho} k_j \alpha_i \alpha_k = \frac{C_{ijkq}}{\rho} k_j \alpha_k \alpha_i \qquad (3.24)$$

and the two expressions on the right-hand side of Equation (3.23) are identical. Hence,

$$\frac{\partial \omega}{\partial k_q} = \frac{\dfrac{C_{ijkq}}{\rho} k_j \alpha_k \alpha_i}{\omega}$$

$$\qquad (3.25)$$

$$S_q = \frac{C_{ijkq} l_j \alpha_k \alpha_i}{\rho \dfrac{\omega}{k}}$$

as desired.
    Now

$$\frac{C_{ijkl}}{\rho} l_j l_l \alpha_k - V^2 \alpha_k \delta_{ik} = 0 \qquad (3.26)$$

Multiplying this equation by α_i yields (since α is a unit vector)

$$V^2 = \frac{C_{ijkl}}{\rho} l_j l_l \alpha_i \alpha_k \tag{3.27}$$

but examining $\boldsymbol{S} \cdot \boldsymbol{l}$ yields

$$\boldsymbol{S} \cdot \boldsymbol{l} = S_l l_l = \frac{C_{ijkl} l_j \alpha_k \alpha_i l_q}{\rho V} = \frac{V^2}{V} = V$$

Hence, the component of the group velocity in the direction of the wave normal is the phase velocity. Therefore, the phase velocity can never be greater than the group velocity.

Energy propagation considerations in anisotropic media yield a somewhat altered view of some fundamental aspects of wave propagation, such as critical angle phenomena. Typically, one approaches critical angle phenomena from the standpoint of Snell's law. Since

$$\frac{\sin \theta_{in}}{V_{in}} = \frac{\sin \theta_{\text{refracted}}}{V_{\text{refracted}}}$$

$$\theta_{\text{refracted}} = \sin^{-1} \left[V_{\text{refracted}} \left(\frac{\sin \theta_{in}}{V_{in}} \right) \right] \tag{3.28}$$

But this requires that

$$V_{\text{refracted}} \frac{\sin \theta_{in}}{V_{in}} \le 1 \tag{3.29}$$

hence, there is an upper limit on θ_{in} for generating a particular wave. However, as pointed out by Henneke [6, 7], Snell's law considerations may be superseded by energy restrictions. For example, the maximum possible θ_{in} predicted from Snell's law may result in the physically unrealizable situation of the energy flux vector for the refracted wave lying outside the boundaries of the refraction media. Hence, one must also examine the orientation of the energy flux vector in determining the maximum incident angle possible.

Also of interest is the observation that, depending on the shape of the slowness surface, the number of transmitted/refracted waves may not be equal to the normal value of three, i.e., one for the quasilongitudinal wave and each of the two possible quasitransverse waves. This

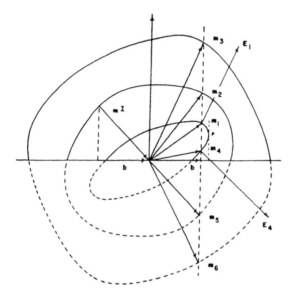

FIGURE 3.2 Effect of the shape of the slowness surface section on the number of reflected and refracted waves.

is illustrated in Figure 3.2. A comprehensive treatment of critical angle in anisotropic media may be found in Reference 7.

As pointed out by Musgrave [8], the normal to the slowness surface must coincide with the direction of the energy flux vector. To see this, consider the equation of the slowness surface

$$M = \det |m_{ik}| = \det |C_{ijkl}m_j m_l - \rho \delta_{ik}| = 0 \qquad (3.30)$$

The normal to the slowness surface will be given by

$$\frac{\partial M}{\partial m_j} = C_{ik} \frac{\partial m_{ik}}{\partial m_j} \qquad (3.31)$$

where C_{ik} is the cofactor of m_{ik}. Since the Christoffel equation is satisfied

$$m_{ik}\alpha_k = 0 \qquad (3.32)$$

and from symmetry

$$M_{ki}\alpha_i = 0 \qquad (3.33)$$

These equations require

$$C_{ik} = f\alpha_i \alpha_k \qquad (3.34)$$

When f is a constant. Therefore,

$$\frac{\partial M}{\partial m_j} \propto C_{ijkl} \alpha_i \alpha_k m_l \qquad (3.35)$$

The normal to the slowness surface must be in the direction of the energy flux vector.

This relationship, in conjunction with the observation that

$$\boldsymbol{S} \cdot \boldsymbol{l} = V \qquad (3.36)$$

allows one to construct the group velocity surface from the slowness surface. Other useful geometric constructions have been developed and are detailed in Musgrave [8].

As the energy flux vector must be normal to the slowness surface, the shape of the slowness surface at the boundary between the incident and refraction media is critical in assessing the importance of these considerations in any critical angle problem. If the slowness surface section for the refracted wave is normal to the boundary at the interface between the two media, then the energy flux vector will coincide with the wave normal and there will be no difference in critical angle as calculated from Snell's law or energy flux considerations. This will be the case for propagation along symmetry axes such as the fiber reinforcement direction (or perpendicular to the fiber reinforcement direction) in a unidirectionally reinforced composite. Rokhlin has exploited this behavior in his work on the experimental determination of elastic moduli in composite media [9]. However, for propagation along an arbitrary (nonsymmetry) direction, this will not be the case and energy flux considerations must be considered for accurate analysis of critical angle phenomena.

CHAPTER 4

Bulk Wave Propagation Anisotropic Media

SAMPLE CALCULATIONS

Example 1—Isotropic Media

For isotropic media, the stiffness matrix is given by

$$\begin{bmatrix} \lambda + 2\mu & \lambda & \lambda & 0 & 0 & 0 \\ \lambda & \lambda + 2\mu & \lambda & 0 & 0 & 0 \\ \lambda & \lambda & \lambda + 2\mu & 0 & 0 & 0 \\ 0 & 0 & 0 & \mu & 0 & 0 \\ 0 & 0 & 0 & 0 & \mu & 0 \\ 0 & 0 & 0 & 0 & 0 & \mu \end{bmatrix} \qquad (4.1)$$

As material properties in an isotropic media are directionally independent, we may choose any direction l for the wave normal and obtain the same result. For convenience, propagation along one of the coordinate system's axes (say the x axis) is chosen so $l = \begin{pmatrix} 1 \\ 0 \\ 0 \end{pmatrix}$. Therefore the equations of motion are satisfied when

$$\det \begin{bmatrix} \lambda + 2\mu - \rho v^2 & 0 & 0 \\ 0 & \mu - \rho v^2 & 0 \\ 0 & 0 & \mu - \rho v^2 \end{bmatrix} = 0 \qquad (4.2)$$

23

The solution to this determinantal equation yields three velocities

$$V_1 = \sqrt{\frac{\lambda + 2\mu}{\rho}}, \qquad V_2 = V_3 = \sqrt{\frac{\mu}{\rho}}$$

One of the corresponding eigenvectors for the initial root is readily found to be

$$\alpha_1 = \begin{pmatrix} 1 \\ 0 \\ 0 \end{pmatrix}$$

Hence, the particle displacements for this wave are parallel to the wave normal as illustrated in Figure 4.1 [10]. The waves are known by a variety of names including longitudinal waves, pressure waves, dilatational waves, and irrotational waves. The remaining two eigenvectors are to a certain extent arbitrary due to the redundant root. Any polarization vector in the plane perpendicular to the propagation direction is possible. Particle motions are illustrated in Figure 4.2 [10]. These waves are known alternatively as shear waves, transverse waves, or equivoluminal waves. Often, the eigenvectors are taken to be

$$\alpha_2 = \begin{pmatrix} 0 \\ 1 \\ 0 \end{pmatrix}, \qquad \alpha_3 = \begin{pmatrix} 0 \\ 0 \\ 1 \end{pmatrix}$$

corresponding to the SH and SV polarizations found in seismology.

If we perform these calculations for all possible propagation directions l and plot the results, the velocity surfaces are obtained. In this case, the surfaces are spherical with two coincident sheets of the velocity surface corresponding to shear wave propagation. A (plane) section of the velocity surface for isotropic media is shown in Figure 4.3. The slowness surface can be calculated directly from the velocity surface. Note that the radius vector coincides with the normal to the surface at every point, hence, the energy flux vector is parallel to the wave normal and the group velocity and phase velocities are equal for every direction of propagation. As was shown in Chapter 3, the direction of energy flux is normal to the tangent line of the slowness surface. Hence, any deviations from a spherical shape will produce energy flux deviation from the wave normal.

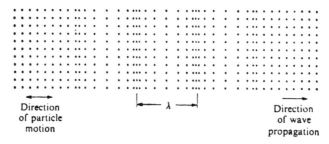

Direction
of particle
motion

λ

Direction
of wave
propagation

FIGURE 4.1 Particle motions for longitudinal waves (after Graff [10]).

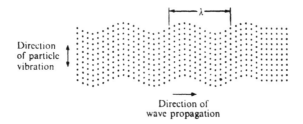

λ

Direction
of particle
vibration

Direction of
wave propagation

FIGURE 4.2 Particle motions for shear waves (after Graff [10]).

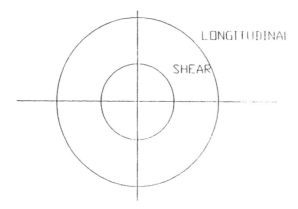

LONGITUDINAL

SHEAR

FIGURE 4.3 Section of velocity surface (isotropic media).

25

For longitudinal wave propagation in the x direction with

$$u = A_o \begin{pmatrix} 1 \\ 0 \\ 0 \end{pmatrix} e^{i(kx - \omega t)}$$

we have

$$\epsilon_{11} = ikA_o e^{i(kx - \omega t)}$$

$$\epsilon_{22} = \epsilon_{33} = \epsilon_{13} = \epsilon_{23} = \epsilon_{12} = 0$$

and

$$\sigma_{11} = (\lambda + 2\mu) ikA_o e^{i(kx - \omega t)}$$

$$\sigma_{22} = \lambda ikA_o e^{i(kx - \omega t)}$$

$$\sigma_{33} = \lambda ikA_o e^{i(kx - \omega t)}$$

$$\sigma_{12} = \sigma_{23} = \sigma_{13} = 0$$

So

$$F = \begin{pmatrix} -(\lambda + 2\mu) k\omega A_o e^{2i(kx - \omega t)} \\ 0 \\ 0 \end{pmatrix}$$

and the energy flux vector coincides with the direction of propagation. Similarly, if we take one of the possible shear polarizations

$$u = A_o \begin{pmatrix} 0 \\ 1 \\ 0 \end{pmatrix} e^{i(kx - \omega t)}$$

we have for the strain field

$$\epsilon_{12} = \frac{1}{2} ikA_o e^{i(kx - \omega t)}$$

$$\epsilon_{11} = \epsilon_{22} = \epsilon_{33} = \epsilon_{13} = \epsilon_{23} = 0$$

So the stress field becomes

$$\sigma_{12} = ik\mu A_o e^{i(kx - \omega t)}$$

$$\sigma_{11} = \sigma_{22} = \sigma_{33} = \sigma_{13} = \sigma_{23} = 0$$

Calculating the energy flux, we have

$$F = \begin{pmatrix} -\mu\omega k A_o^2 e^{2i(kx - \omega t)} \\ 0 \\ 0 \end{pmatrix}$$

which again lies in the direction of the wave normal, as expected from the shape of the slowness surface.

Example 2—Unidirectionally Reinforced Composite (Propagation in a Material Symmetry Direction)

For a unidirectionally reinforced composite, we assume a transversely isotropic structure so

$$C_{ij} = \begin{bmatrix} C_{11} & C_{12} & C_{13} & 0 & 0 & 0 \\ C_{12} & C_{11} & C_{13} & 0 & 0 & 0 \\ C_{13} & C_{13} & C_{33} & 0 & 0 & 0 \\ 0 & 0 & 0 & C_{44} & 0 & 0 \\ 0 & 0 & 0 & 0 & C_{44} & 0 \\ 0 & 0 & 0 & 0 & 0 & \dfrac{C_{11} - C_{12}}{2} \end{bmatrix}$$

where here we have taken the axis of transverse isotropy (i.e., the fiber reinforcement direction) to be the z axis.

$$l = \begin{pmatrix} \sin\theta \\ \cos\theta \\ 0 \end{pmatrix}$$

Then, we have, from the Christoffel equation

$$
\det
\begin{bmatrix}
C_{11} \sin^2 \theta + C_{66} \cos^2 \theta - \rho v^2 & (C_{11} + C_{66}) \sin \theta \cos \theta & 0 \\
(C_{11} + C_{66}) \sin \theta \cos \theta & C_{66} \sin^2 \theta + C_{11} \cos^2 \theta - \rho v^2 & 0 \\
0 & 0 & C_{44} - \rho v^2
\end{bmatrix}
= 0
$$

which, after suitable manipulation, can be factored as

$$
(C_{11} - \rho v^2)(C_{66} - \rho v^2)(C_{44} - \rho v^2) = 0
$$

corresponding to one pure mode longitudinal $(\sqrt{C_{11}/\rho}) = v_1$ and two pure mode shear $(\sqrt{C_{44}/\rho}) = v_2$, $(\sqrt{C_{66}/\rho}) = v_3$ waves. No energy flux deviation is associated with any of these waves. Thus, the situation is quite similar to that found for isotropic media except that the two shear wave speeds are no longer redundant. The slowness surface section appropriate for this case is shown in Figure 4.4. Note that the sections are circular. This is not surprising in light of the fact that the xy plane is a plane of transverse isotropy. Also note that this case reduces to that for isotropy when $C_{44} = C_{66}$. The slowness surface sections will be similar except that the longitudinal curve will lie inside the curve associated with the fast shear wave (shear 1) and the curve for the slower of the shear waves will be outside of the fast wave sheet. There will be no difference between phase and group velocities.

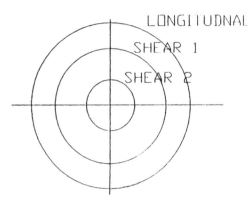

FIGURE 4.4 Velocity surface section for transversely isotropic symmetry (plane of isotropy).

Next, we consider a section (say xz) perpendicular to the plane of transverse isotropy. Now we take the wave normal to be

$$l = \begin{bmatrix} \sin\theta \\ 0 \\ \cos\theta \end{bmatrix}$$

$$u = A_0 \begin{pmatrix} \alpha_1 \\ \alpha_2 \\ \alpha_3 \end{pmatrix} e^{[k(\sin\theta z + \cos\theta y) - \omega t]}$$

Inserting this expression into the equations of motion yields a characteristic equation for the possible velocities of

$$\det \begin{bmatrix} C_{11} \sin^2\theta + C_{44}\cos^2\theta - \rho v^2 & 0 \\ 0 & C_{44}\cos^2\theta + C_{66}\sin^2\theta - \rho v^2 \\ (C_{13} + C_{44})\sin\theta\cos\theta & 0 \end{bmatrix}$$

$$\begin{bmatrix} (C_{13} + C_{44})\sin\theta\cos\theta \\ 0 \\ C_{33}\cos^2\theta + C_{44}\sin^2\theta - \rho v^2 \end{bmatrix} = 0 \quad (4.3)$$

which yields three possible roots

$$V_1 = \sqrt{\left.\frac{-b + \sqrt{b^2 - 4ac}}{2a}\right/\rho}$$

$$V_2 = \sqrt{\left.\frac{-b - \sqrt{b^2 - 4ac}}{2a}\right/\rho} \qquad (4.4)$$

$$V_3 = \sqrt{\frac{C_{44}\cos^2\theta + C_{66}\sin^2\theta}{\rho}}$$

$a = 1$

$b = -[C_{11}\sin^2\theta + C_{44}^2 + C_{33}\cos^2\theta]$

$c = C_{11}C_{44}\sin^4\theta + (C_{44}^2 + C_{11}C_{33} - (C_{13} + C_{44})^2)$

$$\times \sin^2\theta\cos^2\theta + C_{44}C_{33}\cos^2\theta$$

Eigenvectors for these waves may be calculated in the standard fashion. However, only one of the waves yields a convenient polarization vector α. As discussed previously, we had a longitudinal wave when

$$\alpha \cdot l = 1$$

and a transverse wave when

$$\alpha \cdot l = 0$$

These modes are considered to be pure modes, as no transverse component is present in the longitudinal displacement and no longitudinal component is present in the transverse wave displacement. For anisotropic media, it is commonly observed that

$$\alpha \cdot l \neq 0 \qquad \text{or} \qquad 1$$

which makes the classification of the waves somewhat ambiguous as both transverse and longitudinal components of particle displacement will be present. Generally, we classify these waves as quasilongitudinal or quasitransverse, depending on the value of $\alpha \cdot l$ and its proximity to 1 or 0. For this media, the V_3 root yields a pure mode transverse wave with polarization vector

$$\alpha_3 = \begin{pmatrix} 0 \\ 1 \\ 0 \end{pmatrix}$$

However, the remaining two roots correspond to quasilongitudinal (root V_1) and quasitransverse waves (root V_2). Typical velocity and slowness surface sections for a medium with transversely isotropic symmetry are presented in Figure 4.5. Due to the symmetry of the problem, the full velocity and slowness surfaces can be obtained by rotating the sections about the vertical axis in Figures 4.5 (a) and (b), respectively. Of particular interest is the existence of directions where the group velocity is no longer unique due to the cusps on the group velocity curves. This feature can result in some ambiguity in interpreting experimental results when the wave normal has not been firmly established. A comparison of theoretical and predicted experimentally measured slowness surface sections is shown in Figure 4.6 for a unidirectional graphite epoxy sample. Energy flux deviation is found for all three roots. For the pure mode shear and the two nonpure modes, however, the problem is simplified from the most general case in that the energy flux vector will lie in the same plane as the wave normal.

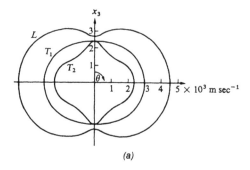

(a)

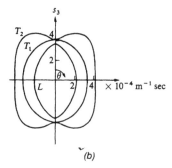

(b)

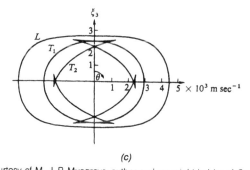

(c)

FIGURE 4.5 Sections for (a) velocity, (b) slowness, and (c) wave surfaces for a medium with transversely isotropic symmetry (after Musgrave [11]).

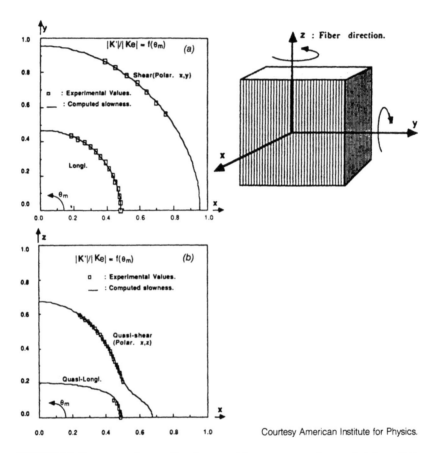

Courtesy American Institute for Physics.

FIGURE 4.6 Slowness surface sections for a graphite-epoxy composite sample (a) xy plane ($\perp$ to fibers) and (b) xz plane. Experimental data is shown for both configurations. Note: The fibers are oriented along the z axis (after Hosten et al. [12]).

Example 3—Orthotropic Media—Symmetry Directions

Next we consider an orthotropic media whose stiffness matrix will be given by

$$C_{ij} = \begin{bmatrix} C_{11} & C_{12} & C_{13} & 0 & 0 & 0 \\ C_{12} & C_{22} & C_{23} & 0 & 0 & 0 \\ C_{13} & C_{23} & C_{33} & 0 & 0 & 0 \\ 0 & 0 & 0 & C_{44} & 0 & 0 \\ 0 & 0 & 0 & 0 & C_{55} & 0 \\ 0 & 0 & 0 & 0 & 0 & C_{66} \end{bmatrix}$$

For wave propagation in one of the three material symmetry planes (say xy), we have for the wave normal

$$l = \begin{pmatrix} \sin \theta \\ \cos \theta \\ 0 \end{pmatrix}$$

and, from the Christoffel equation

$$\det \begin{bmatrix} C_{11} \sin^2 \theta + C_{66} \cos^2 \theta - \rho v^2 & (C_{11} + C_{66}) \sin \theta \cos \theta \\ (C_{11} + C_{66}) \sin \theta \cos \theta & C_{66} \sin^2 \theta + C_{22} \cos^2 \theta - \rho v^2 \\ 0 & 0 \end{bmatrix}$$

$$\begin{matrix} 0 \\ 0 \\ (C_{55} \sin^2 \theta + C_{44} \cos^2 \theta - \rho v^2) \end{matrix} \Bigg] = 0$$

which has the same basic solution character as the transverse isotropy case just considered for a plane perpendicular to the isotropy plane. Again, we find one pure mode transverse wave with a quasilongitudinal wave and a quasitransverse wave as well. Energy flux deviation is observed for all waves as none of the slowness curves are circular. Similar results are obtained for the other planes of material symmetry. Typical results for a sample with orthotropic symmetry are presented in Figures 4.7, 4.8, and 4.9 for the three material symmetry planes. Again, note the presence of cusps in the group velocity surface.

Example 4—Wave Propagation in an Arbitrary Direction in a Composite Lamina

In order to treat the problem of wave propagation in an arbitrary (nonsymmetry) direction in a composite, it is best to consider the simplified problem of a single ply whose fibers are oriented at an arbitrary angle θ with respect to the x axis of our coordinate system. In this way, one may treat each ply in the layup simply by inserting the appropriate angle θ. Note that reflection and refraction phenomena must be considered at each interface between plies.

For a coordinate system rotated, an arbitrary angle θ about the y axis, the new stiffness coefficients (denoted by the ' symbol) are given by

$$C'_{abcd} = a_{ai} a_{bj} a_{ck} a_{dl} C_{ijkl} \tag{4.5}$$

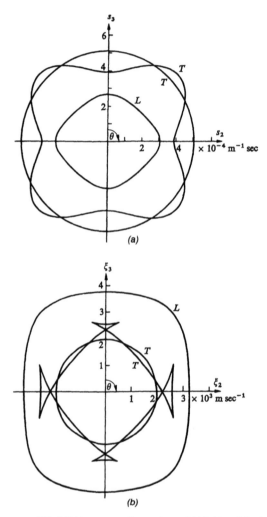

FIGURE 4.7 Representation surfaces for orthotropic media (a) slcwness, (b) group velocity — *yz* plane (after Musgrave [11]).

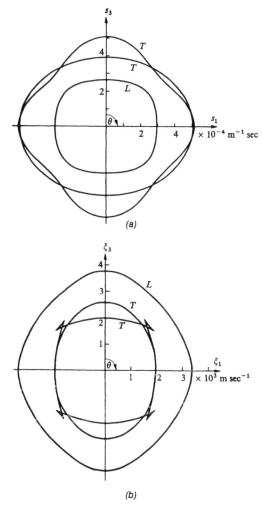

(a)

(b)

FIGURE 4.8 Representation surfaces for orthotropic media (a) slowness, (b) group velocity —xz plane (after Musgrave [11]).

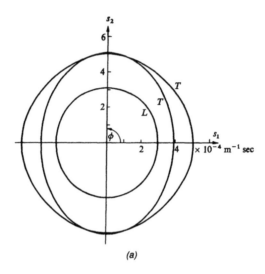

(a)

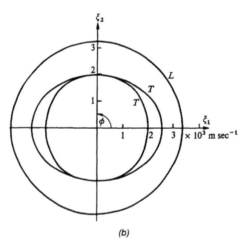

(b)

FIGURE 4.9 Representation surfaces for orthotropic media (a) slowness, (b) group velocity −*xy* plane (after Musgrave [11]).

where

$$\mathbf{q} = \begin{bmatrix} \cos\theta & 0 & \sin\theta \\ 0 & 1 & 0 \\ -\sin\theta & 0 & \cos\theta \end{bmatrix} \qquad (4.6)$$

To simplify the transformation, the reduced notation for the C_{ijkl} is again introduced so the expressions for the primed stiffnesses become:

$$C'_{11} = C_{11}\cos^4\theta + (2C_{13} + 4C_{55})\cos^2\theta\sin^2\theta + C_{33}\sin^4\theta \qquad (4.7a)$$

$$C'_{12} = C_{12}\cos^2\theta + C_{22}\sin^2\theta \qquad (4.7b)$$

$$C'_{13} = C_{12}\cos^2\theta + C_{22}\sin^2\theta \qquad (4.7c)$$

$$C'_{14} = 0 \qquad (4.7d)$$

$$C'_{15} = (C_{13} - C_{11} + 2C_{55})\cos^3\theta\sin\theta$$
$$+ (C_{33} + C_{13} + 2C_{55})\cos\theta\sin^3\theta \qquad (4.7e)$$

$$C'_{16} = 0 \qquad (4.7f)$$

$$C'_{22} = C_{22} \qquad (4.7g)$$

$$C'_{23} = C_{23}\cos^2\theta + C_{12}\sin^2\theta \qquad (4.7h)$$

$$C'_{24} = 0 \qquad (4.7i)$$

$$C'_{25} = (C_{23} - C_{12})\cos\theta\sin\theta \qquad (4.7j)$$

$$C'_{26} = 0 \qquad (4.7k)$$

$$C'_{33} = C_{33}\cos^4\theta + (2C_{13} + 4C_{55})\cos^2\theta\sin^2\theta + C_{11}\sin^4\theta \qquad (4.7l)$$

$$C'_{34} = 0 \qquad (4.7m)$$

$$C'_{35} = (C_{33} - C_{13} - 2C_{55})\cos^3\theta\sin\theta$$
$$+ (C_{13} - C_{11} + 2C_{55})\cos\theta\sin^3\theta \qquad (4.7n)$$

$$C'_{36} = 0 \tag{4.7o}$$

$$C'_{44} = C_{44} \cos^2 \theta + C_{66} \sin^2 \theta \tag{4.7p}$$

$$C'_{45} = 0 \tag{4.7q}$$

$$C'_{46} = (C_{44} - C_{66}) \cos \theta \sin \theta \tag{4.7r}$$

$$C'_{55} = (C_{11} - 2C_{13} + C_{33}) \cos^2 \theta \sin^2 \theta$$

$$+ C_{55}(\cos^2 \theta - \sin^2 \theta) \tag{4.7s}$$

$$C'_{56} = 0 \tag{4.7t}$$

$$C'_{66} = C_{66} \cos^2 \theta + C_{44} \sin^2 \theta \tag{4.7u}$$

For this case, the stiffness matrix is

$$
\begin{bmatrix}
C'_{11} & C'_{12} & C'_{13} & 0 & C'_{15} & 0 \\
C'_{12} & C'_{22} & C'_{23} & 0 & C'_{25} & 0 \\
C'_{13} & C'_{23} & C'_{33} & 0 & C'_{35} & 0 \\
0 & 0 & 0 & C'_{44} & 0 & C'_{46} \\
C'_{15} & C'_{25} & C'_{35} & 0 & C'_{55} & 0 \\
0 & 0 & 0 & C'_{46} & 0 & C'_{66}
\end{bmatrix}
$$

This stiffness matrix represents the mechanical behavior of a lamina where the fibers are oriented at an arbitrary angle θ with respect to a fixed x axis.

For the practical problem of an ultrasonic wave propagating through a given ply in an anisotropic plate, the problem reduces to solving for a wave normal given by

$$
l = \begin{Bmatrix} 0 \\ -\cos \phi \\ \sin \phi \end{Bmatrix}
$$

where ϕ is the angle between the wave normal and the y axis (perpendicular to the plane of lamination). We orient our coordinate system so that the z axis lies in the plane formed by the incident wave normal and the y axis (perpendicular to the laminating plane).

Therefore, we have

$$\lambda_{11} = C'_{66} \cos^2 \phi + C'_{55} \sin^2 \phi \qquad (4.8a)$$

$$\lambda_{12} = -(C'_{25} + C'_{16}) \cos \phi \sin \phi \qquad (4.8b)$$

$$\lambda_{13} = C'_{16} \cos^2 \phi + C'_{25} \sin^2 \phi \qquad (4.8c)$$

$$\lambda_{22} = C'_{22} \cos^2 \phi + C'_{44} \sin^2 \phi \qquad (4.8d)$$

$$\lambda_{23} = -(C'_{23} + C'_{44}) \cos \phi \sin \phi \qquad (4.8e)$$

$$\lambda_{33} = C'_{44} \cos^2 \phi + C'_{33} \sin^2 \phi \qquad (4.8f)$$

$$\boldsymbol{m} = \frac{1}{|\boldsymbol{v}|} \qquad (4.9)$$

Using the slowness vector formulation, the λ' matrix becomes

$$\lambda'_{11} = C'_{66} m_2^2 + C'_{55} m_3^2$$

$$\lambda'_{12} = -(C'_{25} + C'_{46}) m_2 m_3$$

$$\lambda'_{13} = C'_{46} m_2^2 + C'_{35} m_3^2 \qquad (4.10)$$

$$\lambda'_{22} = C'_{22} m_2^2 + C'_{44} m_2^2$$

$$\lambda'_{23} = -(C'_{23} + C'_{44}) m_2 m_3$$

$$\lambda'_{33} = C'_{44} m_2^2 + C'_{33} m_3^2$$

So we seek to find the velocities that make the determinant of the λ' matrix zero, i.e.,

$$\begin{vmatrix} (\lambda'_{11} - \rho) & \lambda'_{12} & \lambda'_{13} \\ \lambda'_{12} & (\lambda'_{22} - \rho) & \lambda'_{23} \\ \lambda'_{13} & \lambda'_{23} & (\lambda'_{33} - \rho) \end{vmatrix} = 0 \qquad (4.11)$$

For this case, the roots usually do not correspond to pure mode propagation and energy flux deviation is commonly observed for all three modes.

CHAPTER 5

Guided Waves

Up to this point, we have considered only bulk wave propagation and have neglected the influence of geometric restrictions on wave propagation. For many cases of practical interest, guided waves can be of great utility in studying the mechanical response of the medium. This is particularly true for composites, which are usually fabricated as plates or platelike structures.

Guided waves have been the subject of a great deal of interest in the seismology and NDE communities since Lord Rayleigh proved that it was possible to propagate an acoustic wave along the surface of a semi-infinite, isotropic solid [13]. A wide variety of guided wave motions have been identified including surface waves, plate (or Lamb) waves [14], Love waves (waves propagating in a thin, solid surface layer overlying a semi-infinite substrate) [15], and Stoneley waves (waves propagating along the boundary of two semi-infinite solids in contact) [16]. The development of the theoretical basis for guided wave propagation has been invaluable in interpreting seismic data. For nondestructive evaluation, however, the uses of guided waves have been significantly restricted, particularly in comparison with bulk waves. This is true even for isotropic media. Guided wave propagation is somewhat more complicated than bulk wave propagation, principally due to the fact that multiple modes of propagation can occur and these modes are usually dispersive. The Rayleigh–Lamb spectrum, which represents the dispersion curves for plate waves in an isotropic medium, is sufficiently complicated (Figure 5.1) without the additional computational difficulty introduced by anisotropy.

Nonetheless, a variety of studies have been conducted to explore the behavior of guided waves (usually surface or plate) in anisotropic media

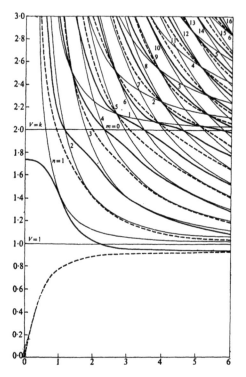

FIGURE 5.1 Dispersion curves for plate wave propagation (isotropic medium, $V = \frac{1}{3}$).

[17–19]. While many aspects of guided wave propagation in composites remain relatively unexplored (particularly Love or Stoneley type waves, and wave propagation in nonsymmetry directions/laminated media), several studies have resulted in significant advances in our understanding of the behavior of guided waves in anisotropic media. One of the principal thrusts in this area has been the development of experimental and analytical techniques to study plates wave propagation in immersion samples from the energy carried away from the sample in the coupling fluid, the so-called "leaky" Lamb wave (LLW) approach.

The general solution for guided waves is very much the same for all types of guided waves of interest in composites. For this discussion, it is advantageous to represent the equations of motion in operator notation. Since we have, from Equation (1.9)

$$C_{ijkl} \frac{\partial^2 u_k}{\partial x_j \partial x_l} = \rho \frac{\partial^2 u_i}{\partial t^2} \qquad (5.1)$$

we may write

$$L_{ik} u_k = \rho \, \frac{\partial^2 u_k}{\partial t^2} \, \delta_{ik} \qquad (5.2)$$

where for anisotropic material (no assumed symmetry)

$$L_{ik} = C_{ijkl} \frac{\partial}{\partial x_j \partial x_l} \qquad (5.3)$$

or

$$L_{11} = C_{11} \frac{\partial^2}{\partial x^2} + C_{66} \frac{\partial^2}{\partial y^2} + C_{55} \frac{\partial^2}{\partial z^2} + 2C_{16} \frac{\partial^2}{\partial x \partial y}$$

$$+ 2C_{15} \frac{\partial^2}{\partial x \partial z} + 2C_{56} \frac{\partial^2}{\partial y \partial z}$$

$$L_{12} = C_{16} \frac{\partial^2}{\partial x^2} + C_{26} \frac{\partial^2}{\partial y^2} + C_{45} \frac{\partial^2}{\partial z^2} + (C_{12} + C_{66}) \frac{\partial^2}{\partial x \partial y}$$

$$+ (C_{14} + C_{56}) \frac{\partial^2}{\partial x \partial z} + (C_{46} + C_{25}) \frac{\partial^2}{\partial y \partial z}$$

$$L_{13} = C_{15} \frac{\partial^2}{\partial x^2} + C_{46} \frac{\partial^2}{\partial y^2} + C_{35} \frac{\partial^2}{\partial z^2} + (C_{14} + C_{56}) \frac{\partial^2}{\partial x \partial y}$$

$$+ (C_{13} + C_{55}) \frac{\partial^2}{\partial x \partial z} + (C_{36} + C_{45}) \frac{\partial^2}{\partial y \partial z}$$

$$L_{22} = C_{66} \frac{\partial^2}{\partial x^2} + C_{22} \frac{\partial^2}{\partial y^2} + C_{44} \frac{\partial^2}{\partial x^2} + 2C_{26} \frac{\partial^2}{\partial x \partial y}$$

$$+ 2C_{46} \frac{\partial^2}{\partial x \partial z} + 2C \frac{\partial^2}{\partial y \partial z}$$

$$L_{23} = C_{56} \frac{\partial^2}{\partial x^2} + C_{24} \frac{\partial^2}{\partial y^2} + C_{34} \frac{\partial^2}{\partial z^2} + (C_{46} + C_{25}) \frac{\partial^2}{\partial x \partial y}$$

$$+ (C_{36} + C_{45}) \frac{\partial^2}{\partial x \partial z} + (C_{23} + C_{44}) \frac{\partial^2}{\partial y \partial z}$$

$$L_{33} = C_{55} \frac{\partial^2}{\partial x^2} + C_{44} \frac{\partial^2}{\partial y^2} + C_{33} \frac{\partial^2}{\partial z^2} + C_{45} \frac{\partial^2}{\partial x \partial y}$$

$$+ C_{35} \frac{\partial^2}{\partial x \partial z} + C_{34} \frac{\partial^2}{\partial y \partial z}$$

Then, we may write

$$\left[L_{ik} - \rho \frac{\partial^2}{\partial t^2} \delta_{ik} \right] u_k = 0 \qquad (5.4)$$

For guided wave problems, we will assume that there is a plane (surface; Stoneley waves, $y = 0$) or set of planes $\left(\text{plate}, y = \pm \frac{t}{2}; \text{Love}, \right.$ $y = 0, t \left. \right)$ which helps to define the motion. Here, we will assume an arbitrary functional dependence on the y coordinate and attempt to find plane wave solutions with an arbitrary propagation direction in the xz plane

$$\boldsymbol{u}(\boldsymbol{x}, t) = A_o \alpha h(y) e^{i(kx\cos\theta + kz\sin\theta - \omega t)} \qquad (5.5)$$

Substituting this solution into Equation (5.4) yields an eigenvalue/eigenvector problem with a non-trivial solution when

$$\det \left[L_{ik} - \rho \frac{\partial^2}{\partial t^2} \delta_{ik} \right] = 0 \qquad (5.6)$$

This determinantal equation yields a sixth-order differential equation for $h(y)$ with solutions of the form $e^{i\beta y}$ so there are six possible values for $\beta = \gamma + i\delta$. The nature of the β's (real, imaginary, complex) govern the character of the solution. For example, for surface waves, we require that the solution remain finite as y gets infinitely far from the surface (say $y \rightarrow + \infty$). Hence, we restrict attention only to those three solutions where the Im $[\beta] = \delta > 0$. Then, the boundary conditions appropriate to the problem are applied. For a free surface, the traction components will be zero, etc. This yields the governing equation for the phase velocity (or wave number) appropriate to the problem.

SURFACE WAVES

For a non-trivial solution to the wave equation

$$\det L_{ij} = 0 \tag{5.7}$$

Assuming solutions of the form

$$u(x, t) = A_o \alpha h(y) e^{i(kx\cos\theta + kz\sin\theta - \omega t)} \tag{5.8}$$

we obtain a sixth-order differential equation for $h(y)$. The solutions are of the form

$$h(y) = e^{i\beta y} \tag{5.9}$$

Hence, there will be six possible values for β. Here, we allow the values for the β's to be complex. This formulation also allows for complex valued displacements via the eigenvector α.

For guided waves, we require that the assumed displacement field satisfy certain boundary conditions appropriate to the particular problem. Surface wave propagation requires that

$$\sigma_{12}|_{y=0} = \sigma_{22}|_{y=0} = \sigma_{23}|_{y=0} = 0 \tag{5.10}$$

since $y = 0$ represents a free surface. In addition, the displacements must remain finite as y increases. Hence, the imaginary part of β must be positive and we may omit contributions from three of the roots. Therefore, the general solution to this problem will be

$$u(x, t) = \sum_{n=1}^{3} A^n \alpha^n e^{-\delta^n y} e^{i(kx\cos\theta + kz\sin\theta + \gamma^n y - \omega t)} \tag{5.11}$$

In addition, the boundary conditions form a set of three linear equations in the unknowns A_i with a zero vector

$$\Delta^* = 0 \tag{5.12}$$

as the solution set. Therefore, there will be a nontrivial solution to this problem if the determinant of the coefficient matrix is zero. This yields the characteristic equation for the wave number, just as for the case of isotropic media. For anisotropic media, however, the determinantal equation is complex; hence, we have

$$\text{Re}\,(\Delta^*) = 0 = \text{Im}\,(\Delta^*) \qquad (5.13)$$

Buchwald has shown that, for most practical situations where the free surface is mirror plane, this equation is purely real [20]. He obtained surface wave solutions for several of these cases. For cases where the free surface did not correspond to a symmetry plane, Buchwald's results indicated directions where surface waves would not propagate. Lim [21] has investigated these so-called "forbidden" directions and found that surface waves would indeed propagate. In his formulation, Lim distinguishes three types of solutions for β, i.e.,

(1) All three β values imaginary
(2) One imaginary β value, two complex
(3) All three β values complex

It should be noted that a complex valued β corresponds to a solution with somewhat different behavior than that observed for the isotropic case, as the decay with distance from the free surface will exhibit both exponential and harmonic characters.

For wave propagation in materials with some degree of internal symmetry, particularly when wave propagation along a symmetry direction is considered, the equations of motion and boundary conditions simplify considerably. The pioneering work in this area is attributable to Stoneley [22] who investigated surface wave propagation in the basal plane of cubic crystals. However, he restricted his attention to solutions with the character of Rayleigh solutions in isotropic media (i.e., solutions that were damped exponentially with depth). This led Stoneley to the conclusion that surface wave propagation was permissible for only a highly restricted class of materials. Synge [23] subsequently showed the existence of permissible solutions that decayed harmonically as well as exponentially with depth and have come to be known as generalized Rayleigh waves.

Consider wave propagation along a symmetry direction (say x) in an orthotropic medium. Following Royer and Dieulesaint [24], we use the following representation

$$u_i = A_o \alpha_i e^{-iqky} e^{i(kx-\omega t)} \qquad (5.14)$$

where

$$q = \beta/k \qquad (5.15)$$

The Christoffel equation then yields

$$\det \begin{bmatrix} (C_{11} + C_{66})q^2 - \rho v^2 & (C_{12} + C_{66})q & 0 \\ (C_{12} + C_{66})q & (C_{66} + C_{22})q^2 - \rho v^2 & 0 \\ 0 & 0 & C_{55} - \rho v^2 \end{bmatrix} = 0 \quad (5.16)$$

If we ignore the pure mode shear wave $v = \sqrt{C_{55}/\rho}$, a nontrivial solution to the Christoffel equation requires

$$C_{22}C_{66}q^4 + [C_{22}(C_{11} - \rho v^2) + C_{66}(C_{66} - \rho v^2)$$

$$- (C_{12} + C_{66})^2]q^2 + (C_{11} - \rho v^2)(C_{66} - \rho v^2) = 0 \quad (5.17)$$

This is a quadratic equation in q^2 which is directly solved using the quadratic formula for the two roots q_1 and q_2. The eigenvectors associated with the two roots are

$$\alpha_1 = \begin{pmatrix} 1 \\ \dfrac{C_{11} - \rho v^2 - C_{66}q_1^2}{(C_{12} + C_{66})q_1} \\ 0 \end{pmatrix} = \begin{pmatrix} 1 \\ p_1 \\ 0 \end{pmatrix}$$

$$(5.18)$$

$$\alpha_2 = \begin{pmatrix} 1 \\ \dfrac{C_{11} - \rho v^2 - C_{66}q_1^2}{(C_{12} + C_{66})q_2} \\ 0 \end{pmatrix} = \begin{pmatrix} 1 \\ p_2 \\ 0 \end{pmatrix}$$

Therefore, the displacements become

$$u_1 = (A_1 e^{-iq_1 ky} + A_2 p_1 e^{-iq_2 ky})e^{i(kx - \omega t)} \qquad (5.19)$$

$$u_2 = (A_1 p_1 e^{-iq_1 ky} + A_2 p_2 e^{-iq_2 ky})e^{i(kx - \omega t)} \qquad (5.20)$$

Substituting these equations into the free surface boundary conditions yields

$$\begin{bmatrix} q_1 + p_1 & q_2 + p_2 \\ (C_{12} + C_{22}q_1p_1) & C_{12} + C_{22}q_2p_2 \end{bmatrix} \begin{pmatrix} A_1 \\ A_2 \end{pmatrix} = 0 \qquad (5.21)$$

which has a nontrivial solution when the determinant is zero. After some algebraic manipulation, the governing equation for Rayleigh wave propagation in this symmetry direction is obtained:

$$C_{22}C_{66}\rho v^2(C_{11} - \rho v^2)$$

$$= (C_{66} - \rho v^2)[C_{22}(C_{11} - \rho v^2) - C_{12}^2]^2 \quad (5.22)$$

The principal difference between these solutions and those for isotropic media is the additional harmonic term in the amplitude decay with depth. If the anisotropy factor A is defined as

$$A = 2C_{66}/C_{11} - C_{12} \qquad (5.23)$$

one can examine the character of the decay in greater detail. Results from Roger and Dieulesaint are presented in Figure 5.2 for five different values of the anisotropy factor. Note that for $A \cong 1$, the character of the solution is much like that for isotropic media. However, as A increases, there is a corresponding increase in the harmonic content of the solution. For surface wave propagation in non-principal directions, the algebra becomes somewhat more complicated but tractable with numerical solution methods.

Rose et al. [25] have studied surface wave propagation in graphite epoxy samples. Variations in phase velocity with propagation direction are presented in Figure 5.3 for unidirectional composites.

PLATE WAVES (FREE)

As was seen in the previous section, the basic governing equations for guided wave propagation simplify considerably for cases where some internal symmetry is present. Here, we proceed in an analogous fashion as before. To illustrate the problem, wave propagation in an arbitrary direction in a transversely isotropic media is considered. The geometry is illustrated in Figure 5.4.

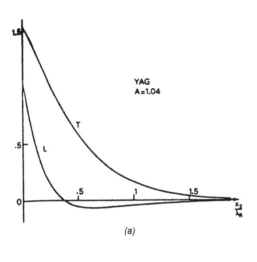

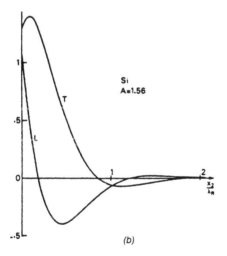

FIGURE 5.2 Rayleigh wave propagation in anisotropic media: Variation of displacement with depth (after Royer and Dieulesaint [24]).

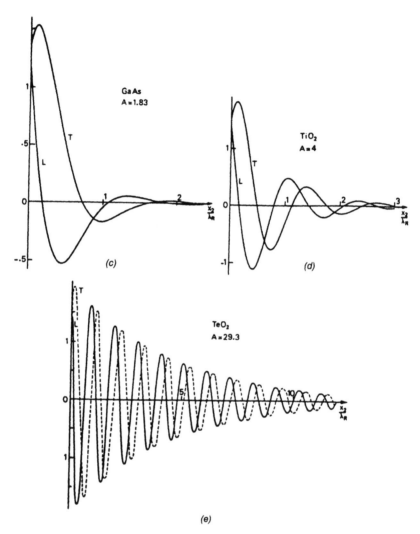

FIGURE 5.2 (continued) Rayleigh wave propagation in anisotropic media: Variation of displacement with depth (after Royer and Dieulesaint [24]).

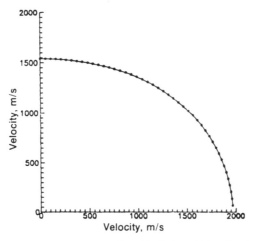

FIGURE 5.3 Variation in phase velocity of surface wave in a unidirectionally reinforced composite (after Rose et al. [25]).

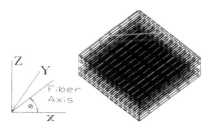

FIGURE 5.4 Geometry for wave propagation in an arbitrary direction in the plane of fiber reinforcement.

51

One of the major simplifications associated with wave propagation in transversely isotropic media is that the equation of motion may be reduced to yield a single relatively simple partial differential equation (the product of a second-order differential operator and a fourth-order differential operator) for each of the displacement components (u, v, w) given by

$$\det L_{ij} = 0 \tag{5.24}$$

where

$$
\det (L_{ij}) = \left[c_{44} \frac{\partial^2}{\partial x^2} + (c_{11} - c_{12}) \left(\frac{\partial^2}{\partial x^2} + \frac{\partial^2}{\partial y^2} \right) - \rho \frac{\partial^2}{\partial t^2} \right]
$$

$$
\times \left[c_{44} c_{23} \frac{\partial^2}{\partial z^2} + \{ c_{44}^2 + c_{11} c_{33} - (c_{13} + c_{44})^2 \} \right.
$$

$$
\times \frac{\partial^2}{\partial z^2} \left(\frac{\partial^2}{\partial x^2} + \frac{\partial^2}{\partial y^2} \right) + c_{44} c_{11} \left(\frac{\partial^2}{\partial x^2} + \frac{\partial^2}{\partial y^2} \right)^2
$$

$$
- \rho(c_{33} + c_{44}) \frac{\partial^4}{\partial z^2 \partial t^2} - \rho(c_{11} + c_{44})
$$

$$
\times \frac{\partial^2}{\partial t^2} \left(\frac{\partial^2}{\partial x^2} + \frac{\partial^2}{\partial y^2} \right) + \rho \frac{\partial^4}{\partial t^4} \right] = 0 \tag{5.25}
$$

For wave propagation in an anisotropic plate, we assume a solution of the form

$$u(x, t) = A_o \alpha h(y) e^{i(kx\cos\theta + kz\sin\theta - \omega t)} \tag{5.26}$$

where

k = wave number
A_o = amplitude
ω = frequency
α = polarization vector

i.e., a plane wave propagating in the plane of the plate at an angle θ with respect to the x axis. Inserting this solution into Equation (5.24), we obtain two differential equations (one second-order, one fourth-order) for $h(y)$ which must be satisfied for wave propagation, i.e.:

$$\Big[[C_{44}\{-k^2 \sin^2 \theta\} h(y) + \frac{1}{2}(C_{11} - C_{12})\{-k^2 \cos^2 \theta h(y)$$

$$+ h''(y)\} + \rho\omega^2 h(y)][c_{44}c_{33}k^4 \sin^4 \theta h(y)$$

$$+ \{c_{44}^2 + c_{11}c_{33} - (c_{13} + c_{44})^2\}\{k^4 \sin^2 \theta \cos^2 \theta h(y)$$

$$- k^2 \sin^2 \theta h''(y)\} + c_{44}c_{11}\{k^2 \cos^2 \theta - 2k \cos \theta h'(y)$$

$$+ h'''(y)\} - (c_{33} + c_{44})\rho\omega^2 k^2 \sin^2 \theta h(y) - \rho\omega^2$$

$$\times (c_{11} + c_{44})\{k^2 \cos^2 \theta h(y) - h''(y)\} + \rho^2\omega^2 h(y)]\Big] = 0 \quad (5.27)$$

where $'$ denotes $\dfrac{\partial}{\partial y}$, $''$ denotes $\dfrac{\partial^2}{\partial y^2}$, etc.

The solutions to these equations are of the form $h(y) = e^{i\beta y}$ which yields three possible values for β^2 (or six values for β), i.e.,

$$\beta_1^2 = \frac{\rho\omega^2 - c_{44}k^2 \sin^2 \theta - \dfrac{1}{2}(c_{11} - c_{12})k^2 \cos^2 \theta}{\dfrac{1}{2}(c_{11} - c_{12})}$$

$$\beta_2^2 = \frac{-b + \sqrt{b^2 - 4ac}}{2a} \qquad\qquad (5.28)$$

$$\beta_3^2 = \frac{-b - \sqrt{b^2 - 4ac}}{2a}$$

where

$$a = c_{11}c_{44}$$

$$b = [c_{44}^2 + c_{11}c_{44} - (c_{13} + c_{44})^2][-k^2 \sin^2 \theta]$$

$$+ 2c_{44}c_{11}(-k^2 \cos^2 \theta) + \rho\omega^2(c_{11} + c_{44})$$

$$c = c_{33}c_{44}k^4 \sin^4 \theta + [c_{44}^2 + c_{11}c_{33} + (c_{13} + c_{44})^2]k^4 \sin^2 \theta \cos^2 \theta$$

$$+ c_{44}c_{11}k^4 \cos^4 \theta - (c_{33}c_{44})\rho\omega^2 k^2 \sin^2 \theta$$

$$+ \rho\omega^2(c_{11} + c_{44})k^2 \cos^2 \theta + \rho^2\omega^4$$

In practice the β^i can be pure real, zero, or pure imaginary, giving rise to different regions of the frequency spectrum.

Once the possible values for β^i have been calculated, the associated polarization vectors (α^i) can then be determined by substituting the potential solution

$$\begin{bmatrix} u \\ v \\ w \end{bmatrix} = A_o^i \begin{bmatrix} \alpha_1^i \\ \alpha_2^i \\ \alpha_3^i \end{bmatrix} e^{i\beta^i y} e^{i(kx\cos\theta + kz\sin\theta - \omega t)} \tag{5.29}$$

into the Christoffel equation and solving for the eigenvectors.

$$\begin{bmatrix} u \\ v \\ w \end{bmatrix} = \sum_{i=1}^{6} A_o^i \begin{bmatrix} \alpha_1^i \\ \alpha_2^i \\ \alpha_3^i \end{bmatrix} e^{i\beta^i y} e^{i(kx\cos\theta + kz\sin\theta - \omega t)} \tag{5.30}$$

The task then is reduced to finding a solution that satisfies the boundary conditions for the problem, which are given by

$$\sigma_{xy}|_{y=\pm b} = 0$$

$$\sigma_{yy}|_{y=\pm b} = 0 \tag{5.31}$$

$$\sigma_{yz}|_{y=\pm b} = 0$$

For a transversely isotropic medium, these equations become

$$\sigma_{xy}|_{y=\pm b} = \left\{\frac{(c_{11}-c_{12})}{2}\right\}\left\{\frac{\partial u}{\partial y}+\frac{\partial v}{\partial x}\right\}\Big|_{y=\pm b}$$

$$= \left\{\frac{(c_{11}-c_{12})}{2}\right\}\sum_{i=1}^{6} A_o^i(\alpha_1^i\beta^i + a_2^i k\cos\theta)e^{\pm i\beta^i b}e^{i\psi} = 0$$

(5.32)

where $\psi = kx\cos\theta + kz\sin\theta - \omega t$

$$\sigma_{yy}|_{y=\pm b} = \left\{c_{12}\frac{\partial u}{\partial x}+c_{11}\frac{\partial v}{\partial y}+c_{13}\frac{\partial w}{\partial z}\right\}\Big|_{y=\pm b} = 0$$

(5.33)

$$\sum_{i=1}^{6}(c_{12}\alpha_1^i k\cos\theta + c_{11}\alpha_2^i\beta^i + c_{13}\alpha_3^i k\sin\theta)e^{\pm i\beta^i b}e^{i\psi} = 0$$

and

$$\sigma_{yz}|_{y=\pm b} = C_{44}\left(\frac{\partial v}{\partial z}+\frac{\partial w}{\partial y}\right)\Big|_{y=\pm b}$$

(5.34)

$$C_{44}\sum_{i=1}^{6} A_o^i(\alpha_2^i k\sin\theta + \alpha_3^i\beta^i)e^{\pm i\beta^i b}e^{i\psi} = 0$$

The task, therefore, is to solve this set of six linear equations in six unknowns. As the solution set for this system of equations is the zero vector, there is a nontrivial solution when the determinant of the coefficient matrix is zero. This yields a characteristic equation for possible wave numbers k for plate wave propagation. Working through the equations, one finds multiple possible modes of propagation analogous to the Rayleigh–Lamb spectrum for isotropic media. Green [26–27] and co-workers have extensively studied the character of plate wave solutions in anisotropic plates. Here, Green's results for the dispersion curve for the lowest order bending mode are presented in Figure 5.5a along with the variations in normal (Figure 5.5b) and shear (Figure 5.5c) stresses for this mode of propagation. Kline et al. [28] have studied plate wave propagation in unidirectionally reinforced composites using a numerical technique to evaluate the results.

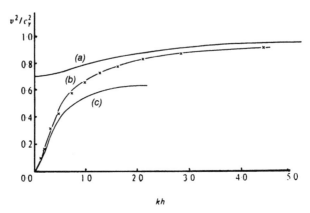

FIGURE 5.5a Dispersion curves for plate waves in (a) idealized inextensible material, (b) transversely isotropic media, (c) approximate solution (after Green [27]).

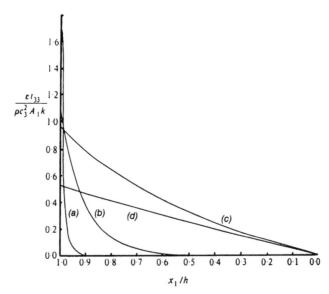

FIGURE 5.5b Variation of normal stress through the thickness of the plate for (a) $kh = 10.322$, (b) $kh = 1.583$, (c) $kh = 0.295$, and (d) $kh = 0.090$ (after Green [27]).

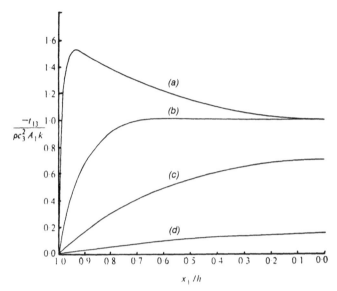

FIGURE 5.5c Variation of shear stress through the thickness of the plate for (a) kh = 10.322, (b) kh = 1.583, (c) kh = 0.295, and (d) kh = 0.090 (after Green [27]).

PLATE WAVES (IMMERSION)

For many practical applications, one is more concerned about the behavior of a plate immersed in a coupling fluid rather than free plate. Accordingly, the bulk of theoretical and experimental work in this area has been directed at fluid loaded plates. The basic form of the wave solution in the plate remains unchanged. The principal differences between the two cases are due to changes in boundary conditions as well as the presence of waves in the fluid which carry energy away from the plate, so-called "leaky" Lamb waves.

Here, we consider a semi-infinite plate bounded on either side by a fluid half-space. For the fluid ($y \geq t/2$ and $\leq -t/2$), we have for the governing equations

$$\bar{\sigma}_{ij,j} = \bar{\rho}\,\ddot{\bar{u}}_i \tag{5.35}$$

and

$$\bar{\sigma}_{ij} = \bar{\lambda}\bar{u}_{k,k}\delta_{ij} \tag{5.36}$$

where the ⁻ denotes quantities appropriate to the fluid. If the plate is isonofied from the upper fluid medium at an angle θ_{in}, the incident wave field will be given by

$$\bar{\boldsymbol{u}}^{in} = \bar{A}^{in} \begin{pmatrix} \sin \theta_{in} \\ \cos \theta_{in} \\ 0 \end{pmatrix} e^{i[\bar{k}(\sin\theta_{in}x + \cos\theta_{in}y - \omega t)]} \qquad (5.37)$$

where

$$\frac{\omega}{\bar{k}} = \bar{V} = \sqrt{\frac{\bar{\lambda}}{\bar{\rho}}} \qquad (5.38)$$

Similarly, the fields for the reflected wave in the upper fluid and the transmitted wave in the lower fluid will be

$$\bar{\boldsymbol{u}}^{re} = \bar{A}^{re} \begin{pmatrix} \sin \theta_{in} \\ -\cos \theta_{in} \\ 0 \end{pmatrix} e^{i[\bar{k}(\sin\theta_{in}x + \cos\theta_{in}y - \omega t)]} \qquad (5.39)$$

and

$$\bar{\boldsymbol{u}}^{t} = \bar{A}^{t} \begin{pmatrix} \sin \theta_{in} \\ -\cos \theta_{in} \\ 0 \end{pmatrix} e^{i[\bar{k}(\sin\theta_{in}x + \cos\theta_{in}y - \omega t)]} \qquad (5.40)$$

The boundary conditions now become

continuity of normal displacement at the fluid-solid interface

$$\bar{u}_2^{in} |_{y=t/2} + \bar{u}_2^{re} |_{y=t/2} = [u_2]_{composite} |_{y=t/2}$$

and

$$\bar{u}_2^{t} |_{y=t/2} = [u_2]_{composite} |_{y=t/2}$$

continuity of normal stress at the fluid-solid interface

$$\bar{\sigma}_{22}^{in} |_{y=t/2} + \bar{\sigma}_{22}^{re} |_{y=t/2} = [\sigma_{22}]_{composite} |_{y=t/2}$$

$$\bar{\sigma}_{22}^{t} |_{y=t/2} = [\sigma_{22}]_{composite} |_{y=t/2}$$

The remaining boundary conditions are vanishing shear stresses at the two interfaces. Therefore, as far as shear is concerned, the interface can be treated as a free surface so these boundary conditions remain unchanged from the earlier treatment

$$[\sigma_{12}]_{\text{composite}} \mid_{y=t/2=0}$$

$$[\sigma_{23}]_{\text{composite}} \mid_{y=t/2=0}$$

$$[\sigma_{12}]_{\text{composite}} \mid_{y=-t/2=0}$$

$$[\sigma_{23}]_{\text{composite}} \mid_{y=-t/2=0}$$

This yields a system of eight equations in eight unknowns which has been studied extensively by a variety of investigators including Chimenti, Nayfeh, Wang, Rokhlin, Mal, Bar-Cohen, and Kinra [29–35] and their co-workers. Nayfeh and Chimenti [29], after considerable algebraic manipulation, have developed closed form solutions for the reflection and transmission coefficients for this problem

$$R = \frac{AS - \lambda}{(S + i\lambda)(A - i\lambda)} \tag{5.41}$$

and

$$Y = \frac{i\lambda(S + A)}{(S + i\lambda)(A - i\lambda)} \tag{5.42}$$

where

$$A = D_{11} D_{23} \tan(\xi d a_1 / a) - D_{13} D_{21} \tan(\xi d a_3 / s)$$
$$S = D_{11} D_{23} \cot(\xi d a_1 / 2) - D_{13} D_{21} \cot(\xi d a_3 / 2)$$
$$Y = \frac{\rho_{\text{fluid}} C^2}{\alpha_f} (W_1 D_{23} - W_3 D_{21})$$
$$\xi = \frac{2\pi f \sin \theta}{v_f}$$
$$D_{1q} = (C_{13} + C_{33} a_q w_q)$$
$$D_{2q} = C_{55}(a_q + w_q)$$
$$w_q = \frac{\rho V^2 - C_{11} - C_{55} a_q}{(C_{13} + C_{55}) a_q}$$

The a_i are solutions to the fourth-order equation

$$Aa^4 + Ba^2 + C = 0 \qquad (5.43)$$

$A = C_{33}C_{55}$

$B = (C_{11} - \rho_s V^2)C_{33} - (C_{55} - \rho_s V^2)C_{55} - (C_{13} + C_{55})^2$

$C = (C_{11} - \rho_s V^2)(C_{55} - \rho_s V^2)$

and

$$a_f = \frac{V^2}{V_f^2} - 1$$

Typical dispersion curve results are shown in Figure 5.6.

LOVE TYPE WAVES IN COATED ANISOTROPIC MEDIA

The propagation of guided waves in coated materials, or Love waves, has also been of some interest for anisotropic media. The most comprehensive treatment of the problem can be found in the work of Farnell and Adler [37]. They examined the general problem of wave propagation in thin layers and considered the case of a thin, anisotropic layer overlaying an anisotropic substrate in some detail. For composite materials, we are particularly concerned about this particular type of guided wave for the inspection of composites that require a coating layer for environmental protection. One of the principal applications of such coating is for carbon-carbon composites that require an oxidation protection layer (such as silicon-carbide) to prevent excessive mass loss during prolonged exposure to elevated temperatures. While Farnell and Adler [37] were concerned with wave propagation in material symmetry directions, more recently Bouden and Datta [38] have reexamined the problem for situations appropriate to composites. In this work, the case of guided wave propagation in a transversely isotropic medium with an isotropic coating was studied. Wave propagation at arbitrary angles with respect to the fiber axis was considered.

The general formulation of the problem has much in common with the guided wave solutions previously considered. As with surface waves, a solution is sought which decays with distance into the substrate. Now,

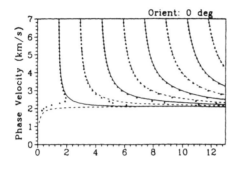

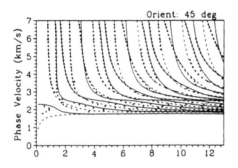

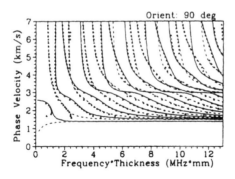

FIGURE 5.6 Dispersion curves for plate wave propagation in a unidirectionally reinforced composite at 0°, 45°, 90° with respect to the fiber reinforcement (after Mal et al. [31]).

however, one must consider the influence of a surface layer. For both the substrate and coating, admissible solutions are of the form

$$u(x, t) = A_o h(y)\alpha e^{i(kx\cos\theta + kz\sin\theta - \omega t)} \qquad (5.44)$$

where $h(y)$ is of the form $e^{i\beta y}$ with the six possible values of β determined from the wave equation. Again we let $\beta^n = \gamma^n + i\delta^n$ as for surface waves, the requirement of finite displacements for positions far removed from the surface require that the imaginary part of the β's be positive. Hence, three of the roots may be omitted from the substrate solution to yield

$$u_{\text{sub}}(x, t) = \sum_{n=1}^{3} A^n \alpha^n e^{\delta^n y} e^{i(kx\cos\theta + kz\sin\theta + \gamma^n y - \omega t)} \qquad \delta^n > 0 \quad (5.45)$$

For the coating layer, all of the possible β values must be retained in the solution, i.e.,

$$u_{\text{coat}}(x, t) = \sum_{n=1}^{9} A^n \alpha^n e^{\delta^n y} e^{i(kx\cos\theta + kz\sin\theta + \gamma^n y - \omega t)} \qquad (5.46)$$

The nine boundary conditions required for solution of the problem arise from the requirements to zero surface traction on the free surface $y = 0$

$$(\sigma_{12})_{\text{coat}}|_{y=0} = (\sigma_{22})_{\text{coat}}|_{y=0} = (\sigma_{23})_{\text{coat}}|_{y=0} = 0 \qquad (5.47)$$

and continuity of traction and particle displacement at the interface between the two media

$$u_{\text{coat}}|_{y=H} = u_{\text{sub}}|_{y=H}$$

$$v_{\text{coat}}|_{y=H} = v_{\text{sub}}|_{y=H}$$

$$w_{\text{coat}}|_{y=H} = w_{\text{sub}}|_{y=H}$$

$$(\sigma_{12})_{\text{coat}}|_{y=H} = (\sigma_{12})_{\text{sub}}|_{y=H}$$

$$(\sigma_{22})_{\text{coat}}|_{y=H} = (\sigma_{22})_{\text{sub}}|_{y=H}$$

$$(\sigma_{23})_{\text{coat}}|_{y=H} = (\sigma_{23})_{\text{coat}}|_{y=H}$$

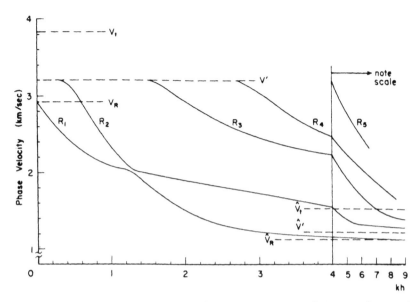

FIGURE 5.7 Dispersion curves for Rayleigh modes (gold layer on nickel substrate, after Farnell and Adler [37]).

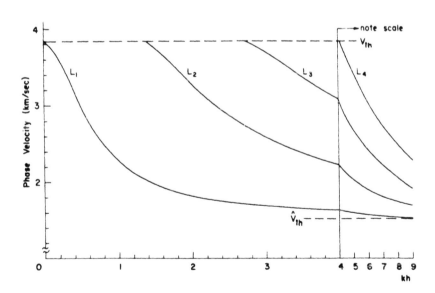

FIGURE 5.8 Dispersion curves for Love modes (gold layer on nickel substrate, after Farnell and Adler [37]).

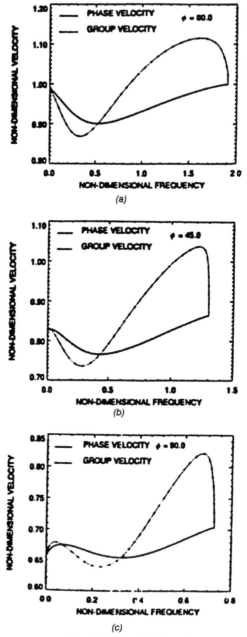

FIGURE 5.9 Dispersion curves for Love wave propagation (aluminum layer on graphite-epoxy, after Bouden and Datta [38]). (a) Direction of propagation along the fiber axis; (b) direction of propagation rotated 45° from the fiber axis; and (c) direction of propagation normal to fiber axis.

The dispersion curves associated with wave propagation in an isotropic medium with an isotropic coating layer can often be grouped into two distinct families—Rayleigh modes and Love modes. However, for anisotropic substrates, this grouping is not always possible as coupled modes are observed. Figures 5.7 and 5.8 illustrate uncoupled modes for the case of an aligned anisotropic layer (gold) on an anisotropic substrate (nickel); both with cubic symmetry. The direction of propagation is along a common symmetry axis and energy propagates in the direction of the wave normal. Note that for the limiting case of a very thin surface layer relative to the wavelength, the lowest-order Rayleigh mode velocity degenerates into the Rayleigh velocity expected for the substrate. Similarly, for the lowest-order Love mode, the limiting velocity is the shear velocity (horizontal polarization) of the substrate. For fiber-reinforced media, the results to date have been quite limited. Bouden and Datta have identified one "Rayleigh-like" mode that will propagate for a limited range of possible wave lengths for isotropic aluminum on transversely isotropic graphite epoxy. The low frequency limit, as before, is the Rayleigh velocity of the composite and the high frequency limit is the lowest transverse (quasitransverse) velocity of the composite substrate. This behavior is illustrated in Figure 5.9, where normalized dispersion curves are presented for this mode for propagation, with wave normal, parallel, perpendicular, and at 45° with respect to the fiber axis. Note the difference between group and phase velocities, even for propagation along symmetry axes (0° and 90°). Clearly, more work is required in this area to identify useful modes of Love wave propagation appropriate for composite applications.

CHAPTER 6

Experimental Consideration for Ultrasonic Measurements

As was shown earlier, the velocity of wave propagation in an arbitrary direction in a composite will be a function of the direction cosines of the wave normal and the elastic moduli of the media. Hence, ultrasonic velocity measurements can in principle be used to nondestructively measure elastic constants. This fact is well appreciated in isotropic media where velocity measurements are commonly used to replace destructive mechanical tests to accomplish this purpose. For composite media, however, several difficulties have slowed this application of ultrasonic testing. One of the principal difficulties encountered is the algebraic complexity of the problem. In isotropic media, the problem of extracting pertinent moduli is relatively simple as

$$V_L = \sqrt{\frac{\lambda + 2\mu}{\rho}}, \qquad V_S = \sqrt{\frac{\mu}{\rho}}$$

Hence

$$\mu = \rho V_S^2$$

and

$$\lambda = \rho V_L^2 - 2\rho V_S^2$$

Here, we are dealing with only two moduli, and the equations are linear in V_S^2 and V_L^2. For composites, this is no longer the case. Here, we are forced to deal with additional constants (five for transversely isotropic

67

media, nine for orthotropic media, etc.) and in many cases nonlinear equations. This places increased demands on experimental accuracy and data processing capabilities.

TIME DELAY MEASUREMENTS (BULK WAVES)

The precise determination of material properties via ultrasonic testing hinges upon the accurate measurement of transit time. Direct visual observations of echo traces on an oscilloscope are usually insufficient for this purpose, and therefore a variety of more sensitive techniques have been devised. It should be pointed out that in this case we are dealing primarily with absolute velocity measurements rather than measuring relative velocity changes from point to point within the medium. This, of course, is a somewhat more difficult task.

Prior to the advent of high-speed waveform digitizers, the majority of applications were analog based. Two of the more popular methods for accurate analog velocity measurement are the sing-around method and the pulse-echo overlap method. In the sing-around method, resolution enhancement is achieved through the use of multiple echoes in the time delay characterization. For pulse-echo testing, the first returned echo is used as a trigger source for a new test pulse. With this approach, the repetition rate for the transducer excitation is the parameter of interest, as it is directly related to the round trip transit time in the sample. An electronic counter is used for this purpose. A block diagram for this system is shown in Figure 6.1. Truell et al. [39] estimate the sensitivity to be approximately one part in 5×10^5 with the sing-around approach.

A second commonly used approach, the pulse-echo overlap method, utilizes (in one embodiment) an adjustable pulse repetition rate to produce a superposition of successive echoes in the echo train. An electronic counter is then used to measure the repetition rate, and hence the transit time in the sample. A schematic diagram for such a system is shown in Figure 6.2. Accuracy with this approach is typically reported to be on the order of one part in 10^4 although higher values may be obtained under specialized circumstances [40]. It should be pointed out that other experimental factors such as transducer coupling, parallelism of reflecting surfaces, and diffraction effects of the transducer, etc, can significantly compromise the accuracy of these measurements.

More recently, digital-based signal processing techniques have supplanted analog methods for most ultrasonic transit time measurements. The principal limitation in accurate digitally based, time delay mea-

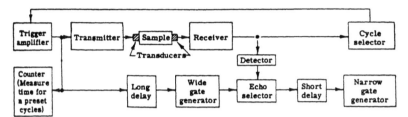

FIGURE 6.1 Experimental schematic for the sing-around method (after Truell et al. [39]).

surements is the inherent speed of the digitizer, typically 100 MHz or one sample every 10 nsec. To improve upon this intrinsic time resolution, one of several signal processing techniques can be used. Many investigators utilize either auto or cross correlation algorithms for this purpose [41, 42] depending on whether pulse-echo or through transmission geometries are used. If we assure that the reference signal (say the first echo) given by $R(t)$ and the sample signal (say the second echo) or $S(t)$ are identical except for a time delay τ_{in}, [i.e., $S(t) = R(t) - \tau_1$] the correlation function will be given by

$$C_{RS}(\tau) = \int_{-\infty}^{+\infty} R(t)S(t - \tau)dt$$

$$= \int_{-\infty}^{+\infty} R(t)R[t - (\tau + \tau_1)]dt \quad .$$

Naturally, the maximum in the correlation function will occur when the two signals coincide or $\tau = -\tau_1$. In practice, due to the oscillatory nature of the transducer pulses generated, there are several local maxima

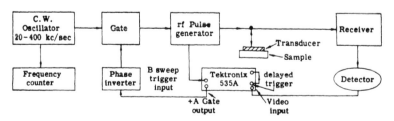

FIGURE 6.2 Experimental schematic-pulse echo overlap (after Truell et al. [39]).

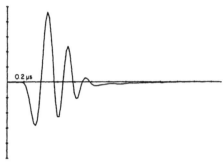

Reference signal for a 5 MHz central frequency transducer

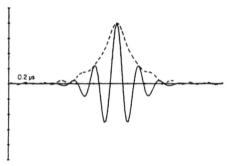

Autocorrelation function

FIGURE 6.3 Use of correlation analysis for time delay measurements (after Castagnede et al. [42]).

in the correlation spectrum. The actual time delay τ_1 is determined from the maximum in the correlation envelope, as illustrated in Figure 6.3. As the correlation function is calculated from digital data, it is limited by the base sample time of the digitizing unit, as before. However, as a maximum in the envelope is sought, by interpolating between points a much improved time delay measurement may be obtained. Egle [41] estimates this improvement to be a factor of 100 over the basic sample rate of the digitizer. Typical results from a composite sample are shown in the waveform of Figure 6.4, where the quasilongitudinal and quasitransverse modes are both excited. Clearly, one can discover both echoes in this case. However, as the signal arrivals come closer together in time, the maximum resolution capability of this approach is reached as illustrated in Figure 6.5 for signals with simulated time delays of $40t$, $25t$, and $15t$ where t is the single interval of the digitizer.

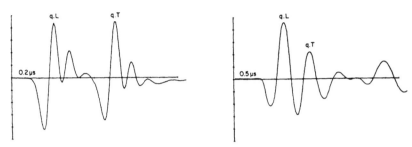

FIGURE 6.4 Use of correlation analysis to separate QL & QT modes of propagation in composites (after Castagnede et al. [42]).

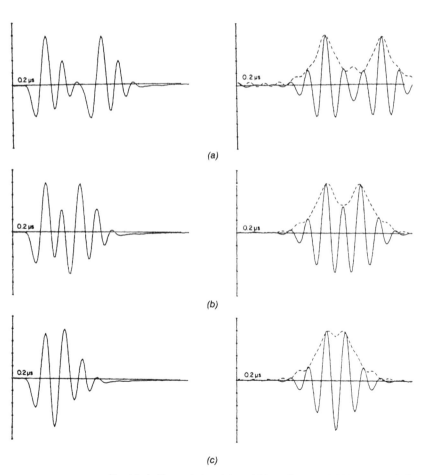

FIGURE 6.5 Effect of echo separation on resolution (after Castagnede et al. [42]).

71

Fourier techniques can also be used for a similar purpose and with the use of FFT algorithms are quite effective. Chang and co-workers [43] have shown that the time delay between two successive ultrasonic echoes may be determined from the power spectrum of the ultrasonic signals. To see this, consider two identical signals separated by a time delay τ [$f(t)$ and $f(t + \tau)$]. The Fourier transform of the two signals will be

$$\mathcal{F}[f(t)] = \frac{1}{\sqrt{2\pi}} \int_{-\infty}^{+\infty} e^{i\omega t} f(t)\,dt = \bar{f}(\omega)$$

and

$$\mathcal{F}[f(t + \tau)] = \frac{1}{\sqrt{2\pi}} \int_{-\infty}^{+\infty} e^{i\omega t} f(t + \tau)\,dt$$

which, via a change of variable $u = t + \tau$ becomes

$$= \frac{1}{\sqrt{2\pi}} \int_{-\infty}^{+\infty} e^{i\omega(u-\tau)} f(u)\,du$$

$$= \bar{e}^{i\omega\tau} \bar{f}(\omega)$$

or for the combined signal

$$\mathcal{F}[f(t) + f(t + \tau)] = (1 + e^{-i\omega\tau})\bar{f}(\omega)$$

Therefore, the Fourier transform of the combined signal will be the reference spectrum $f(\omega)$ modified by a factor of $1 + e^{-i\omega\tau}$. The time delay τ can be directly determined from the anti-resonance minima in the power spectrum as

$$|1 + e^{-i\omega}|^2 = |(1 + \cos \omega\tau) + i \sin \omega\tau|^2 = 2 + 2 \cos \omega\tau$$

which has minima at π, 3π, 5π, etc. Experimentally, this is most easily accomplished by using the front surface reflection (pulse-echo) or the direct water path (through-transmission) signals as reference for de-convolving the transducers characteristics [$\bar{f}(\omega)$] from the total response. Usually, this approach is used for normal incidence. However, Chang et al. [43] have shown that the same principle can be adapted to oblique incidence (immersion) for wavespeed determination via the relationship

$$C_S = \frac{2TC_0\Delta f}{\sqrt{4T^2\Delta f^2 \sin \theta_{in}^2 + C_o^2}}$$

where

C_S = sample velocity
C_0 = velocity in coupling fluid
T = specimen thickness
Δf = period of minima in amplitude spectra
θ_{in} = angle of incidence

The process is illustrated in Figures 6.6–6.8 for normal and oblique incidence.

Kinra and co-workers [44, 45] have used a variation of this Fourier transform based approach to determine the wavespeed in thin materials. In this work, they use both the magnitude and phase spectra to extend the procedure into the fractional wavelength domain. Typical results are shown in Figure 6.9 where the sensitivity of both the magnitude

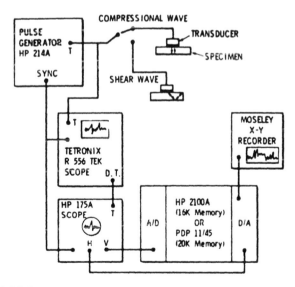

FIGURE 6.6 Schematic for ultrasonic resonance technique (after Chang et al. [43]).

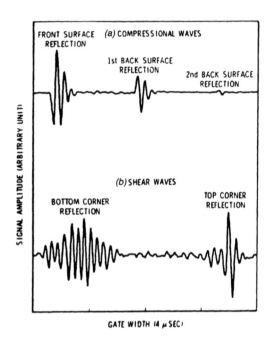

FIGURE 6.7 Waveforms for normal (longitudinal) oblique (shear) waves (after Chang et al. [43]).

and phase spectra are shown for a 0.31 mm thick graphite-epoxy specimen immersed in water and a transducer center frequency of 2.25 MHz. They found that the spectra are sensitive to small changes in wavespeed near specimen resonances where $2h/\lambda = m$, $m = 1$, 2, 3, etc.

Other techniques have also been developed for resolving closely spaced echoes in thin samples. Heyser [46] has shown that the square of the magnitude of the analytic signal is proportional to the rate of arrival of one of the components of energy, which may be kinetic energy, potential energy, or a linear combination of both. The square of the real signal will thus be zero at any instant when one of its component energies is zero, whereas the square of the analytic signal magnitude will only be zero when the total instantaneous energy is zero. Since analytic signal magnitude is directly related to the rate of energy arrival, it is the optimal estimator of interface location for echo signals commonly used in ultrasonic analysis. Gamel [47] has used this as the basis for ultrasonic velocity measurements in thin samples.

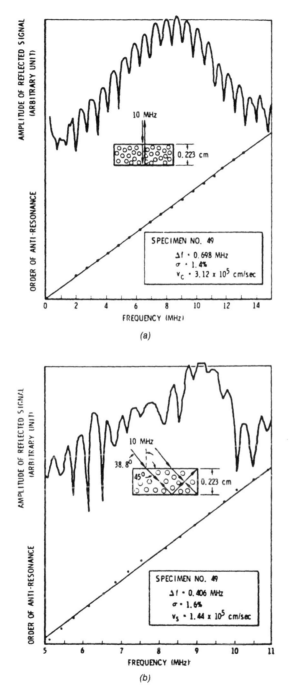

FIGURE 6.8 Antiresonance spectra. (a) normal incidence; (b) oblique incidence (after Chang et al. [43]).

75

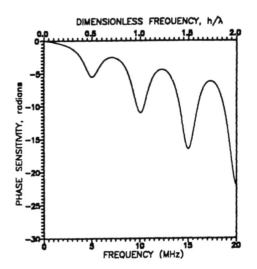

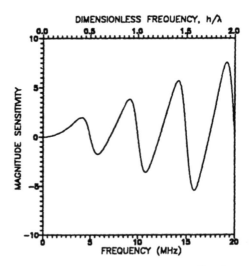

FIGURE 6.9 Sensitivity of magnitude and phase spectra to velocity changes after Iyer et al. [45]).

When treating an ultrasonic signal as an analytic signal, the real signal is replaced by its complex form. For a simple harmonic function, the real signal is represented by

$$f(t) = a \cos \omega t + b \sin \omega t$$

This is replaced by

$$h(t) = f(t) + ig(t)$$

Here, a and b are constants. The function $g(t)$ is obtained from $f(t)$ by replacing $\cos \omega t$ by $\sin \omega t$ and $\sin \omega t$ by $-\cos \omega t$.

The conventional rectification of the ultrasonic signal is replaced by computing the magnitude of the analytic signal. The detection scheme uses the magnitude of the analytic signal instead of rectification, and produces a signal that is proportional to the square root of the rate of arrival of energy at the transducer. If the real part of the signal is recorded, then the imaginary part $V_i(t)$ can be obtained by the Hilbert transform as given below.

$$V_i(t) = \int_{-\infty}^{\infty} V_r(t) \pi^{-1} (t - t^*)^{-1} dt^*$$

This equation represents a convolution of the signal with the kernel $1/(\pi t)$. By the convolution theorem, a convolution of two functions can be implemented in the frequency domain by the multiplication of their Fourier transforms. Therefore, the convolution can be performed in the frequency domain by multiplying the Fourier transform of $V_r(t)$ by $-i \, \text{sgn}(f)$ (where $\text{sgn}(f) = -1$ for $f < 0$, $= 1$ for $f > = 0$) and then applying an inverse Fourier transform to the product. The capabilities of this approach are demonstrated in Figure 6.10 for a single layer of graphite-epoxy prepreg material.

While the bulk of experimental interest in composite materials characterization is in absolute rather than relative velocity measurement, relative measurements can be quite useful, particularly for scanning large parts—providing an accurate absolute velocity measurement can be made at a base point in the sample. One of the most powerful approaches to this measurement test employs automatic frequency control to accurately track changes from point to point within the sample. With this approach the pulse width of the transducer excitation is adjusted until the echoes overlap within the sample. The interference

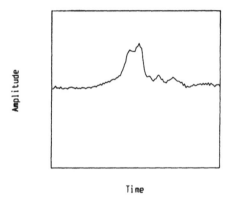

FIGURE 6.10 Analytical signal magnitude separation of front and back surface echoes in composite prepreg (oblique incidence).

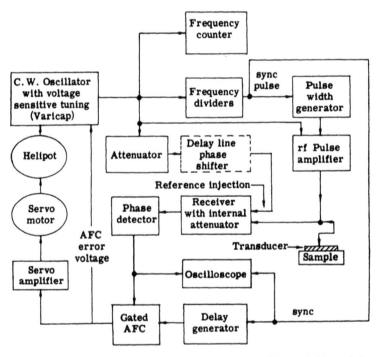

FIGURE 6.11 Schematic of pulse phase locked loop system (after Truell et al. [39]).

between the various echoes produces a series of maxima and minima corresponding to constructive and destructive interference. By varying the frequency of the source, the interference pattern will be changed and the velocity obtained from

$$\frac{\Delta V}{V} = \frac{\Delta f}{f}$$

where

Δf = frequency separation between adjacent frequency producing the same interference pattern

For scanning purposes, one operates from a null point and adjusts (via AFC) the driving source frequency to maintain the interference null. By carefully tracking the frequency changes with a frequency counter, one obtains a good measure of the relative velocity changes from point to point within the specimen. A schematic diagram of such a phase locked loop system is shown in Figure 6.11. This type of system requires an accurate absolute velocity measurement at the initial point of the scan, as only relative changes are monitored. However, with this proviso, this approach is sufficiently rapid to permit the integration of full scale parts.

GUIDED WAVE MEASUREMENTS

The experiments used to characterize guided wave modes differ substantially from those used for bulk wave measurements in that measurements over a wide band width are needed to fully characterize the dispersion relationship for the various guided wave modes which can propagate in composites. These tests are usually performed in immersion, with efforts focused on the energy radiated back into the coupling medium, the so-called "leaky" waves. While "leaky" modes may be found for a variety of guided waves, they have been found to be most useful for plate (or Lamb) wave propagation in composites, hence the term "Leaky" Lamb waves or LLW. For dispersion curve measurements, one uses a pitch-catch arrangement with sensing and receiving transducers oriented at a pre-selected angle ($\pm\theta$) with respect to the surface normal. The incident wave interacts with the test sample and produces a specular reflection as well as the leaky wave field. The specular reflection and leaky wave field interface to produce the two-field structure separated by a null zone, as shown in Figure 6.12. This in-

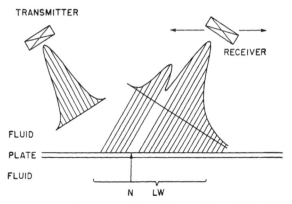

FIGURE 6.12 Leaky Lamb wave acoustic field (after Chimenti and Martin [49]).

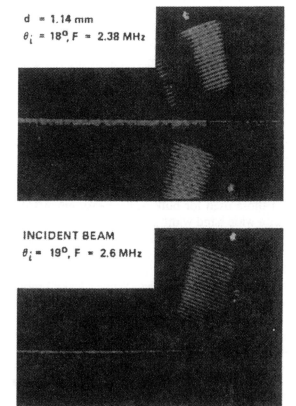

$d = 1.14$ mm

$\theta_i = 18°, F = 2.38$ MHz

INCIDENT BEAM

$\theta_i = 19°, F = 2.6$ MHz

FIGURE 6.13 Schlieren photograph of leaky Lamb wave field (after Bar-Cohen [48]).

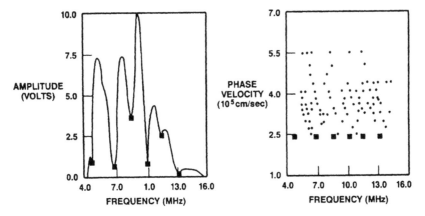

FIGURE 6.14 Leaky Lamb wave data acquisition (after Bar-Cohen [48]).

terference phenomenon can also be visualized as in the Schlieren photographs in Figure 6.13. From Snell's law considerations, the coupling of the incident acoustic wave field with the test medium is most efficient when the projection of the incident wave number on the specimen surface is equal to the wave number of the guided wave of interest

$$\frac{K_{in}}{\sin \theta_{in}} = K_{\text{guided}}$$

The null zone is produced by destructive interference in the reflected wave field and is a very sensitive feature that can be used to characterize the various modes of guided wave propagation.

Since guided wave propagation is usually dispersive, it is desirable to conduct measurements over a wide frequency range. One way to accomplish this is to use a tone burst excitation and vary the frequency over a wide bandwidth. Results from this type of experiment are presented in Figure 6.14. Alternatively, a wide band signal may be generated and analyzed using Fourier transform techniques. Note the presence of minima in the spectrum. These minima correspond to destructive interference between the specular surface reflection and particular guided wave modes radiated back into the coupling media. For the simple case of the generating and receiving transducers aligned with the fiber axis, Nayfeh and Chimenti [29] give the reflection coefficient as

$$R = \frac{AS - Y^2}{(S + iY)(S - iY)}$$

where A, S, and Y are defined in Equation (5.42). The minima correspond to the case when a particular guided mode is excited and

$$AS = Y^2$$

which can be solved for the corresponding wave number K_{guided} (hence phase velocity since $V_{phase} = \omega/K$) for that mode. Since the frequency of the tone burst is known, this yields a point in the dispersion curve. Successive minima produce additional points in the spectrum. By repeating this experiment over a wide range of incident angles, a full picture of the dispersion relations can be obtained. Several alternative procedures for reconstructing these dispersion curves using spectral analysis techniques have been developed to improve the speed and accuracy of the data acquisition and analysis and are described in detail in References [48–50].

CHAPTER 7

Methods for Elastic Modulus Reconstruction from Ultrasonic Data

As presented in Chapters 2 and 3, there is a fundamental relationship between ultrasonic phase (or group) velocity and mechanical properties, which can be used to nondestructively determine anisotropic elastic moduli. For materials with orthotropic (or higher) symmetry, shear and longitudinal wave propagation in symmetry directions can be used to directly yield the diagonal components of the stiffness matrix

$$\bar{C} = \rho\bar{v}^2$$

Off diagonal components require a prior knowledge of the diagonal component, propagation in a non-symmetry direction (usually via sectioning) and a little more algebraic manipulation. However, for most composite structures, sectioning techniques are not possible and the plate geometry precludes accurate in-plane velocity measurements for full characterization of the diagonal components. Therefore, alternatives must be found.

One approach to the problem in composite plates is to directly measure the through thickness stiffness (e.g., C_{22} in the geometry shown in Figure 7.1) from the normal incidence longitudinal velocity and use oblique incidence at multiple angles of incidence and numerical techniques to solve the resulting system of nonlinear equations for the remaining moduli. For orthotropic materials (in the coordinate system shown), the governing Christoffel equation takes the following form. Here we consider phase velocity measurements in a plane of material symmetry for illustration, as the algebra is significantly easier to deal with than that of the general case.

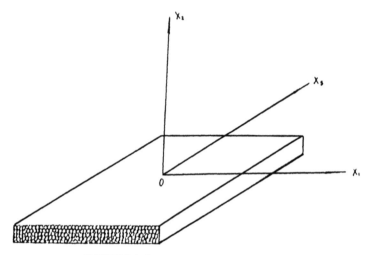

FIGURE 7.1 Orthotropic specimen geometry.

$$
\det \begin{bmatrix}
C_{66} \cos^2 \theta + C_{55} \sin^2 \theta - \rho v^2 & 0 \\
0 & C_{22} \cos^2 \theta + C_{44} \sin^2 \theta - \rho v^2 \\
0 & (C_{23} + C_{44}) \sin \theta \cos \theta
\end{bmatrix}
$$

$$
\begin{bmatrix}
0 \\
0 \\
C_{44} \cos^2 \theta + C_{33} \sin^2 \theta - \rho v^2
\end{bmatrix} = 0
$$

Since the pure mode transverse wave

$$
\rho v^2 = C_{66} \cos^2 \theta + C_{55} \sin^2 \theta
$$

cannot be excited, using conventional immersion techniques, the roots of interest are the quasilongitudinal and quasitransverse velocities given

$$
\rho v_{QL}^2 = \frac{-b + \sqrt{b^2 - 4C}}{2}
$$

and

$$
\rho v_{QT}^2 = \frac{-b - \sqrt{b^2 - 4C}}{2}
$$

where

$$b = -(C_{44} \cos^2 \theta + C_{55} + C_{33} \sin^2 \theta)$$

and

$$C = (C_{22} \cos^2 \theta + C_{44} \sin^2 \theta)(C_{44} \cos^2 \theta + C_{33} \cos^2 \theta)$$

$$- (C_{23} + C_{44})^2 \sin^2 \theta \cos^2 \theta$$

The value of C_{22} is known from the initial measurement at normal incidence. The task, then, is to measure three independent velocities and solve for the unknowns C_{44}, C_{23}, and C_{33}. This can be achieved at three distinct propagation angles, say θ_1, θ_2, and θ_3, from three different incidence angles in the immersion medium. To avoid ambiguity in identifying a particular mode, one usually concentrates on the initial arrival (*QL* below the first critical angle and *QT* above.)

Therefore, we now have three coupled nonlinear equations:

$$\text{at each } \theta_i \qquad f_i(C_{44}, C_{23}, C_{33}) = [\Delta T]_{\text{predicted}} - [\Delta T]_{\text{measured}} = 0$$

This formulation can be readily modified to group velocity measurements instead of phase velocity via Equation (3.25). However, as we shall show in Chapter 8, while phase and group velocity measurements do not coincide, for many practical experimental geometries, phase velocity can be directly measured using a suitable signal for relative time delay measurements.

The solution of coupled, nonlinear systems of equations is one of the more challenging problems in numerical analysis. The existence of multiple local minima often makes finding an actual root difficult. Usually, one employs an iterative approach to the problem. However, this approach requires some approximate knowledge of the solution so that an initial guess can be made. Then, one searches a range of values in the neighborhood of this guess for the actual root of the system of equations. Of particular importance in this search process are the speed and accuracy of the convergence process.

The Newton–Raphson method (or Newton's method) and its variants are among the most widely used approaches to the problem [51]. The basis of this method can be easily illustrated geometrically in one dimension as follows. Suppose we wish to find the root of a nonlinear function of a single variable $f(x)$ in the vicinity of the x_0 as shown in Figure 7.2. Since the line tangent to the curve $f(x)$ is a reasonable

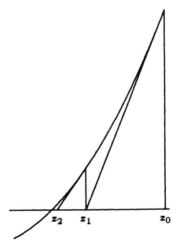

FIGURE 7.2 Illustration of Newton–Raphson (one variable, after Elden and Wittmeyer–Koch [51]).

approximation of the behavior in the region, it can be used to find an improved estimate (x_1) of the actual root. The equation of the tangent line is given by

$$y = f(x_0) + f'(x_0)[x - x_0]$$

Then, the new estimate of the root will be given by

$$x_1 = x_0 - \frac{f(x_0)}{f'(x_0)}$$

The process is repeated using the recursion relationship

$$x_{n+1} = x_n - \frac{f(x_n)}{f'(x_n)} \qquad n = 0, 1, 2, \cdots$$

until convergence is obtained. The Newton–Raphson method will always converge to the proper root if the initial approximation is sufficiently close to the actual root. This is equivalent to using a Taylor series to approximate functional behavior in a region, say

$$f(x_n + h_n) = f(x_n) + f'(x_0)h_n + \text{H.O.T.} = 0$$

So the desired correction h_n to the present estimate x_n is given by

$$h_n = \frac{-f(x_n)}{f'(x_n)}$$

as before.

The approach for nonlinear systems of equations is again similar with a Taylor series expansion used for approximation. For a nonlinear system

$$f_i(x_1, x_2, \cdots, x_m) = 0$$
$$\vdots$$
$$f_m(x_1, x_2, \cdots, x_m) = 0$$

or in vector form

$$\boldsymbol{f}(\boldsymbol{x}) = 0$$

For each component of $\boldsymbol{f}$, we have for the nth iterate

$$f_i(\boldsymbol{x}^n + \boldsymbol{h}^n) = f_i(\boldsymbol{x}^n) + \sum_{j=1}^{m} \frac{\partial f_i}{\partial f_j}(\boldsymbol{x}^n) \cdot h_j^n + \text{H.O.T.} \quad i = 1, 2, \cdots, m$$

Denoting the Jacobian of the function by J, this expansion can be rewritten as

$$\boldsymbol{f}(\boldsymbol{x}^n + \boldsymbol{h}^n) = \boldsymbol{f}(\boldsymbol{x}^n) + J(\boldsymbol{x}^n)\boldsymbol{h} + \cdots$$

Solving for the root, we have for the correction term

$$\boldsymbol{h}^n = -J(\boldsymbol{x}^n)^{-1}\boldsymbol{f}(\boldsymbol{x}^n)$$

so

$$\boldsymbol{x}^{n+1} = \boldsymbol{x}^n + \boldsymbol{h}^n$$

Another iterative approach to the problem is based on the observation that the system of equations

$$\boldsymbol{f}(\boldsymbol{x}) = 0$$

is solved whenever the function

$$S(x) - f_1^2(x) + f_2^2(x) + f_3^2(x) + \cdots + f_m^2(x)$$

is minimized. Again, one begins with an initial estimate $x^0 = (x_1^0, x_2^0, \cdots, x_m^0)$. The gradient vector $\nabla S(x^0)$ is then calculated to determine the direction of steepest descent to the minimum. The next approximation is then given by

$$x' = x^0 - t\nabla S(x^0)$$

The value of t is usually chosen to minimize S in this direction. Steepest descent methods can be combined with other approaches to form hybrid methods such as the Marquardt algorithm [52] which employs both the Newton–Raphson and steepest descent approaches.

Since many of the available algorithms for the solution of nonlinear systems of equations are iterative, requiring the use of an initial guess, the question of convergence must be raised. One must take care that the restrictions on the accuracy of the initial guess are not so restrictive as to preclude convergence when noisy data with experimental uncertainty is used. Kline et al. [53] have studied sensitivity to this initial estimate for the case of transverse isotropy. Synthetic data with a known solution were utilized with the initial moduli guesses and were varied in a systematic fashion. Typical results (two-parameter variation) are shown in Figure 7.3, where the diamond, ♦, indicates the actual value of the solution, and the region of convergence is shaded. Note that when one of the parameters is very close to the actual value, considerably greater than normal uncertainty in the remaining variables can be tolerated. Nonetheless, it is clear that convergence is obtained over a sufficiently large range (typically $\pm 50\%$) to allow one to conclude that this uncertainty does not present a serious problem for most practical situations.

A further area of considerable interest is that of the sensitivity of ultrasonic velocity measurements to all pertinent elastic moduli, particularly the in-plane moduli. Wooh and Daniel [54] have recently studied this aspect of the problem in detail for the case of transverse isotropy. For moduli determined from direct wavespeed measurements, uncertainty in modulus is simply twice that of the uncertainty in the basic velocity measurement as for these moduli

$$C_{ij} = \rho v^2$$

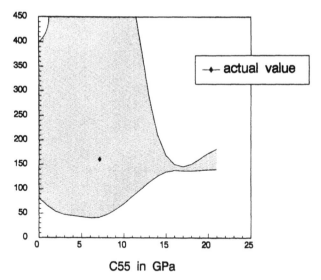

FIGURE 7.3 Convergence region for elastic modulus reconstruction for varying initial guesses (after Kline et al. [53]).

hence

$$\frac{\Delta C_{ij}}{C_{ij}} = 2\,\frac{\Delta v}{v}$$

For other moduli, the situation is considerably more complicated. First of all, many of these calculations are based on previous measurements. In the case just cited, the determination of C_{44}, C_{23}, and C_{33} was based in part on the accuracy of the C_{22} measurement obtained directly from the through thickness velocity measurement. Hence, errors will be propagated in successive calculations. Further, consider the eigenvalue problem of the form posed earlier

$$\det \begin{bmatrix} \lambda_{11} - \rho v^2 & \lambda_{13} \\ \lambda_{13} & \lambda_{33} - \rho v^2 \end{bmatrix} = 0$$

so

$$\lambda_{13}^2 = (\lambda_{11} - \rho v^2)(\lambda_{33} - \rho v^2)$$

Here, Wooh and Daniel point out that the most critical term is the

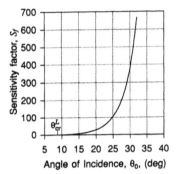

FIGURE 7.4 Sensitivity analysis results for Poisson's ratio (after Wooh and Daniel [54]).

$(\lambda_{33} - \rho v^2)$ term, which they denote by A. For oblique incidence in immersion, we may write (using Snell's law)

$$A = \lambda_{33} - \rho v^2 = n_1^2 \left(C_{55} - C_{33} - \frac{\rho v_w^2}{\sin^2 \theta_0} \right) + C_{33}$$

where

v_w = sound velocity in water
θ_0 = incidence angle
n_1 = cosine of the angle of refraction

Then the sensitivity is given by

$$\frac{\partial A}{A} = \left[\frac{n_1}{A} \frac{\partial A}{\partial n_1} \right] \frac{\Delta n_1}{n_1} = S_f \frac{\Delta n_1}{n_1}$$

Figure 7.4 shows the sensitivity of this measurement to the choice of incidence angle θ_0. From this study, Wooh and Daniel recommend that this angle be chosen as close as possible to the critical angle for optimal results.

One viable solution to the inherent problem of the insensitivity of a particular velocity measurement to a particular elastic modulus is to oversample and take more measurements than minimally required for property reconstruction. In this way, the effects of random errors introduced by experimental error will be minimized, as they will tend to cancel out each other. In the solution of an overdetermined system of equations, one may employ least squares minimization procedures such as the steepest descent method described earlier, except that now the system of equations may be inconsistent, i.e., there may not be one actual solution due to experimental errors. Here, one seeks to minimize the sum of the least square errors without expecting to find an actual root. This essentially amounts to the curve fitting of the slowness surface, dispersion curve, etc. to the experimental data with redundancy introduced by the oversampling. Many of the recent papers in elastic modulus reconstruction employ this approach, and may be expected to have greater accuracy in modulus determination than techniques that are minimally sampled. Also, there may be an additional advantage to using overdetermined systems as they free one from having to take optimal advantage of material symmetry.

For orthotropic plates, for example, instead of taking measurements along the two symmetry axes in the plane of the plate to isolate the contributions from the various moduli (essentially reducing the problem to two uncoupled sets of three nonlinear equations in three unknowns and finding the remaining moduli in a nonsymmetry plane, usually at 45° with respect to the symmetry axes in the plane of fiber reinforcement); totally arbitrary directions may be employed. Mignogna and co-workers [55] have achieved good success in materials with symmetry through the use of twenty-seven measurements in nonsymmetry directions to solve for the nine unknowns using a least squares minimization procedure.

Recently, Karim et al. [56] have employed an alternative algorithm to the least squares minimization approach to curve fitting. Their approach is based on the work of Caceci and Cacheris [57] and employs a simplex algorithm to achieve this end. A simplex is a polygonal geometric figure whose number of vertices are one larger than the dimensions of the space under consideration (i.e., a triangle in two-

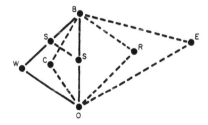

FIGURE 7.5 Simplex model (after Caceci and Cacheris [57]).

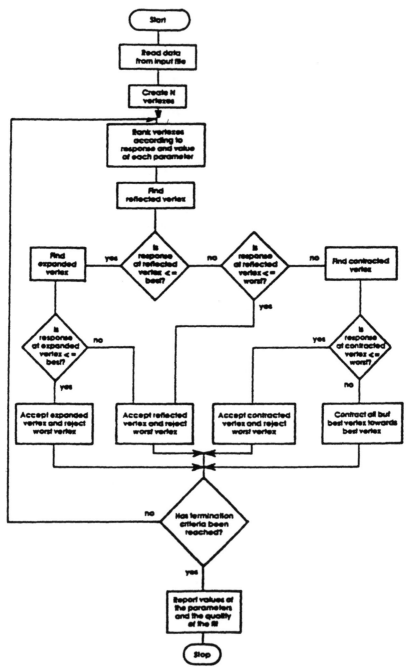

Start

Read data
from input file

Create N
vertexes

Rank vertexes
according to
response and value
of each parameter

Find
reflected vertex

Find
expanded
vertex

Is
response
of reflected
vertex < =
best?

Is
response
of reflected
vertex < =
worst?

Find contracted
vertex

Is
response
of expanded
vertex < =
best?

Is
response
of contracted
vertex < =
worst?

Accept expanded
vertex and reject
worst vertex

Accept reflected
vertex and reject
worst vertex

Accept contracted
vertex and reject
worst vertex

Contract all but
best vertex towards
best vertex

Has termination
criteria been
reached?

Report values of
the parameters
and the quality
of the fit

Stop

yes   no   no

no   yes

yes   no

yes   no

no

yes

FIGURE 7.6 Flowchart for simplex algorithm (after Caceci and Cacheris [57]).

dimensional space, tetrahedron in three-dimensional space, etc.) Each vertex then represents a possible set of solution values to the problem. The response at each vertex is determined and evaluated on the basis of the sum of the squares of the errors in each measurement. The algorithm then identifies the vertices with the largest (W) and smallest (B) errors and replaces the point with the worst fit with an alternative chosen in a systematic fashion. Vertex positions are adjusted via the mechanisms of reflection, expansion, contraction, and shrinkage as illustrated in the two-dimensional space model (simplex BWO) in Figure 7.5. The process is then repeated iteratively until convergence is achieved. The basic simplex algorithm is presented in Figure 7.6. This approach may have a computational advantage over conventional methods, principally in that it avoids numerical differentiation and the attendant round off errors.

At this stage of development, there is no single, generally accepted approach to the problem of elastic property reconstruction. However, a variety of methods are being developed to improve the speed and accuracy of nonlinear equation solving in a noisy (due to experimental error) environment. Certainly, the chances of finding an acceptable solution are improved when uncertainties in transducer positioning and time delay measurement are kept to a minimum and when redundant measurements are made. It is also clear that despite the problems inherent in any nonlinear system, virtually all of the available approaches yield satisfactory results without overly stringent experimental requirements.

CHAPTER 8

Experimental Considerations for Dynamic Modulus Measurement in Anisotropic Media: Phase Velocity vs. Group Velocity

As we have seen, one of the difficulties that recurs in elastic moduli reconstruction stems from the difference between phase and group velocity for an anisotropic medium. For modulus reconstruction purposes, the governing equations are for the most part conveniently expressed in terms of phase velocity, yet experimentally one measures group velocity. This can have serious consequences for geometries such as the point source/point sensor technique, where there may be some difficulty in establishing the direction of the wave normal. Hence, other experimental configurations such as the line source approach may be easier to implement.

One can modify the governing nonlinear equations from phase to group velocity using the relations found previously in Equation (3.25)

$$S_k = \frac{C_{ijkl}l_j\alpha_i\alpha_j}{\rho V_{\text{phase}}}$$

where

$$V_{\text{group}} = \text{mag}(S)$$

and

$$S \cdot l = V_{\text{phase}} = |S||l|\cos\bar{\psi}$$

However, there are certain experimental configurations that allow direct phase velocity calculation from time delay measurements, eliminating the need for this type of modification.

95

AS4/3501-6 [0]

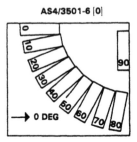

FIGURE 8.1 Geometry of sectioned samples for velocity measurements (after Pearson and Murri [58]).

The problem was recently addressed by Pearson and Murri [58] who considered the problem for two cases—constant measurements and immersion measurements (symmetry plane) for a transversely isotropic medium. First, they sectioned samples at 10° increments and measured the through transmission wave speed as shown in Figure 8.1. Then they calculated the group velocity deviation angle $\bar{\psi}$ and computed the normal stiffness in the direction of the wave normal where $\bar{\psi}$ was (a) properly accounted for and (b) ignored. The results of this study are shown in Figure 8.2. Note the discrepancy in the calculated stiffness, particularly for angles removed from the symmetry directions (0° and 90°

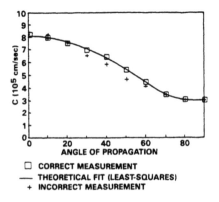

□ **CORRECT MEASUREMENT**
—— **THEORETICAL FIT (LEAST-SQUARES)**
+ **INCORRECT MEASUREMENT**

FIGURE 8.2 Illustration of error introduced in failing to properly account for energy flux deviation from wave normal (after Pearson and Murri [58]).

show no difference, as for these two angles no deviation is observed). They also demonstrated that, for through transmission geometries and propagation in certain symmetry directions where S and l and the surface normal V all lie in a common plane, the phase velocity may be directly calculated from the measured time delay, provided one uses the water path delay (no sample between the two transducers) as a reference. These authors correctly point out that this is the principal reason why several early investigators were able to obtain correct moduli despite the fact that energy flux deviation was ignored. In a later work, Mignogna [59] showed that the Pearson and Murri results could be extended to samples of any arbitrary symmetry. Kline and co-workers [60] further generalized the problem to include pulse-echo configurations. In this section, a thorough discussion of the experimental geometries that permit the direct calculation of phase velocity from the measured time delays will be presented. Therefore, these configurations will probably be the geometries of choice for elastic modulus reconstruction, as group velocity corrections will not be needed.

NORMAL INCIDENCE/CONTACT MEASUREMENTS

One of the most direct ways in which phase velocity information can be directly extracted from the delay measurements is via a through transmission experiment, as illustrated in Figure 8.3. Here, one measures transit time Δt as

$$\Delta t = \frac{l^*}{V_{group}}$$

Here, with a contact transducer, we have a clearly defined wave normal, perpendicular to the transducer/sample interface. Since we have already seen that the projection of the group velocity along the wave normal is, in fact, the phase velocity

$$S \cdot l = V_{phase}$$

Hence

$$V_{group} \cos \psi = V_{phase}$$

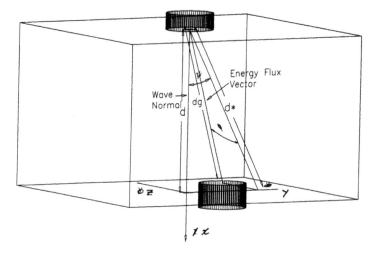

FIGURE 8.3 Experimental geometry for contact measurements: pulse-echo and through transmission.

and

$$\Delta t = \frac{l/\cos \psi}{V_{\text{phase}}/\cos \psi} = \frac{l}{V_{\text{phase}}}$$

This is precisely what one would have obtained if energy flux deviation was totally ignored, providing that the transducer was moved from its usual position (or was sufficiently large) to pick up the signal on the back surface. This result is valid for all materials, regardless of internal symmetry.

For pulse-echo geometries, a contact (or normal incidence immersion) transducer placed on the surface ($x_2 x_3$ plane) of a generally anisotropic medium (Figure 8.3) also applies if the back surface transducer is ignored. We are particularly interested in the relation between the energy flux vector associated with the incident wave on the back surface $\mathbf{S}^{in}$ and the energy flux vector associated with the reflected wave from the back surface $\mathbf{S}^{re}$. For this case, the incident and reflected wave normals will be given by

$$\mathit{l}^{in} = \begin{pmatrix} 1 \\ 0 \\ 0 \end{pmatrix} \quad \text{and} \quad \mathit{l}^{re} = \begin{pmatrix} -1 \\ 0 \\ 0 \end{pmatrix}$$

The Christoffel equation for this problem assumes the simple form

$$
\begin{pmatrix}
C_{11} - \rho v^2 & C_{16} & C_{15} \\
C_{16} & C_{66} - \rho v^2 & C_{56} \\
C_{15} & C_{56} & C_{55} - \rho v^2
\end{pmatrix}
\begin{pmatrix}
\alpha_1 \\
\alpha_2 \\
\alpha_3
\end{pmatrix} = 0
$$

for both the incident and reflected waves. Hence, the polarizations for the two waves must be the same, i.e.,

$$
\alpha^{in} = \alpha^{re} = \alpha
$$

Now, we may calculate the components of the energy flux vectors using the relations

$$
S_i = \frac{C_{ijlm}\alpha_j\alpha_m l_l}{\rho V}
$$

as

$$
S_i^{in} = \frac{C_{ijlm}\alpha_j\alpha_m}{\rho V} \quad \text{and} \quad -\frac{C_{ijlm}\alpha_j\alpha_m}{\rho V} = S_i^{re}
$$

Thus, we may conclude that $S^{in} = -S^{re}$, which requires that the path of energy propagation for the reflected wave will be precisely the same as that followed by the incident wave. Therefore, just as for the through transmission experiment,

$$
\Delta t = \frac{2l^*}{V_{group}} = \frac{2l/\cos\psi}{V_{phase}/\cos\psi} = \frac{2l}{V_{plane}}
$$

and the phase velocity can be directly recovered. The principal advantage of this geometry is that now we don't even need to worry about transducer positioning, as the energy is returned directly to the generating transducer.

OBLIQUE INCIDENCE

For oblique incidence, it is readily found that the phase velocity can be directly recovered from time delay measurements for typical through transmission or pulse-echo geometries, provided one uses the proper water path signal as a reference for the time delay measurements.

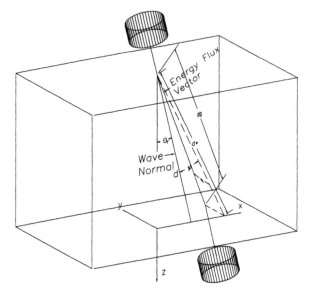

FIGURE 8.4 Experimental geometry for immersion measurements: through transmission.

Through Transmission

The typical geometry for a through transmission experiment is shown in Figure 8.4. Remember that Snell's law requires that the incident and refracted wave normals lie in a common plane, but the direction for the refracted energy flux does not usually lie in this plane. In general, the energy flux vector will lie in a cone about the wave normal. Here, we may write

$$\mathbf{l}_i = \begin{pmatrix} 0 \\ \cos \theta_i \\ \sin \theta_i \end{pmatrix} \tag{8.1}$$

$$\mathbf{l}_r = \begin{pmatrix} 0 \\ \cos \theta_r \\ \sin \theta_r \end{pmatrix} \tag{8.2}$$

and

$$\mathbf{S} = V_{\text{group}} \begin{pmatrix} \sin \phi \\ \cos \phi \cos (\theta_r + \psi) \\ \cos \phi \sin (\theta_r + \psi) \end{pmatrix} \tag{8.3}$$

where, from Snell's law

$$\frac{\sin \theta_i}{V_w} = \frac{\sin \theta_r}{V_{\text{phase}}} \tag{8.4}$$

and from our previous discussion about the relationship between phase and group velocities

$$\mathbf{S} \cdot \mathbf{l}_{re} = V_{\text{phase}} \tag{8.5}$$

This requires that

$$V_{\text{group}} \cos \phi \cos \psi = V_{\text{phase}} \tag{8.6}$$

Here, we use ϕ to represent the angle between $\mathbf{S}$ and the plane formed by $\mathbf{l}^{in}$ and $\mathbf{l}^{re}$ and ψ to represent the angle between the projection of $\mathbf{S}$ onto this plane and $\mathbf{l}^{re}$.

The measured time delay between the signals observed both with and without the sample inserted in the beam will be given by

$$\Delta t = t_{w\,/\text{o sample}} - t_{w\,\text{sample}}$$

$$= \frac{d_w}{V_w} - \frac{dg}{V_g} \tag{8.7}$$

Working with the projection of the group velocity ray, $dg \cos \phi$, which forms an angle of $\theta_r + \psi$ with the normal to the surface, we have

$$dg \cos \phi \cos (\theta_r + \psi) = d \tag{8.8}$$

Similarly, working with the right triangle formed by dw and the projection of the group velocity ray $dg \cos \phi$, we have

$$dw = dg \cos (\theta_r - \theta_i + \psi) \tag{8.9}$$

Therefore, the expression for the time delay becomes

$$\Delta t = \frac{dg \cos \phi \cos (\theta_r - \theta_i + \psi)}{V_w} - \frac{dg}{V_g} \tag{8.10}$$

Substituting for dg [Equation (8.8)] and using Equation (8.6) yields

$$\Delta t = \frac{d}{\cos(\theta_r + \psi)} \left[\frac{\cos(\theta_r - \theta_r + \psi)}{V_w} - \frac{\cos\psi}{V_{phase}} \right] \tag{8.11}$$

Expanding the $\cos(\theta_r - \theta_i + \psi)$ term

$$\Delta t = \frac{d}{\cos(\theta_r + \psi)}$$

$$\times \left[\frac{\cos(\theta_r + \psi)\cos\theta_i}{V_w} + \frac{\sin(\theta_r + \psi)\sin\theta_i}{V_w} - \frac{\cos\psi}{V_{phase}} \right] \tag{8.12}$$

Again, applying Snell's law [Equation (8.4)] to the middle term

$$\Delta t = \frac{d}{\cos(\theta_r + \psi)}$$

$$\times \left[\frac{\cos(\theta_r + \psi)\cos\theta_i}{V_w} + \frac{\sin(\theta_r + \psi)\sin\theta_r - \cos\psi}{V_{phase}} \right] \tag{8.13}$$

Expanding the $\sin(\theta_r + \psi)$ term and multiplying $\cos\psi$ by 1 (as $\cos^2\theta_r + \sin^2\theta_r$) yields

$$\Delta t = \frac{d}{\cos(\theta_r + \psi)} \left[\frac{\cos(\theta_r + \psi)\cos\theta_i}{V_w} - \frac{\cos(\theta_r + \psi)\cos\theta_r}{V_{phase}} \right] \tag{8.14}$$

$$\Delta t = d \left[\frac{\cos\theta_i}{V_w} - \frac{\cos\theta_r}{V_{phase}} \right] \tag{8.15}$$

Letting $\cos\theta_r = \sqrt{1 - \sin^2\theta_r}$ and solving for V_{phase} yields

$$V_{phase} = \left[\left(\frac{\Delta t}{d} \right)^2 - \frac{2\Delta t/d}{V_w}\cos\theta_i + \left(\frac{1}{V_w} \right)^2 \right]^{-1/2} \tag{8.16}$$

which is precisely the same expression obtained for isotropic media using phase velocity considerations alone. Therefore, even though the measured quantity Δt reflects energy propagation with the group velocity, the phase velocity may be found directly from the time delay measurements (just as the isotropic case), providing that the direct water path transit time is used as a reference. Again, this result is valid for any anisotropic medium regardless of material symmetry.

Pulse-Echo

For pulse-echo geometries, the considerations are much the same as for through transmission. Here, the front surface reflection is used as the reference for the time delay measurements. The wave normals for the problem will be

$$l_i = \begin{pmatrix} 0 \\ \cos \theta_i \\ \sin \theta_i \end{pmatrix}, \qquad l_r = \begin{pmatrix} 0 \\ \cos \theta_r \\ \sin \theta_r \end{pmatrix} \quad \text{as before} \qquad (8.17)$$

and

$$l_{re} = \begin{pmatrix} 0 \\ -\cos \theta_r \\ \sin \theta_r \end{pmatrix}$$

for the reflected wave from the back surface.

For this case, the Christoffel elements become

$$\lambda_{11} = C_{66} \cos^2 \theta_r + C_{55} \sin^2 \theta_r \pm 2 C_{56} \sin \theta_r \cos \theta_r$$

$$\lambda_{22} = C_{22} \cos^2 \theta_r + C_{44} \sin^2 \theta_r \pm 2 C_{24} \sin \theta_r \cos \theta_r$$

$$\lambda_{33} = C_{44} \cos^2 \theta_r + C_{33} \sin^2 \theta_r \pm 2 C_{34} \sin \theta_r \cos \theta_r$$

$$(8.18)$$

$$\lambda_{23} = C_{24} \cos^2 \theta_r + C_{34} \sin^2 \theta_r \pm \frac{1}{2} (C_{23} + C_{44}) \sin \theta_r \cos \theta_r$$

$$\lambda_{13} = C_{46} \cos^2 \theta_r + C_{35} \sin^2 \theta_r \pm \frac{1}{2} (C_{36} + C_{45}) \sin \theta_r \cos \theta_r$$

$$\lambda_{12} = C_{26} \cos^2 \theta_r + C_{45} \sin^2 \theta_r \pm \frac{1}{2} (C_{25} + C_{46}) \sin \theta_r \cos \theta_r$$

The complicating effect of the change of sign in the y component of the wave normal makes generalizations such as those obtained for the through transmission case difficult. If, however, we assume that the xz plane is a plane of material symmetry, a simplified result may be obtained that will be valid for most materials of practical interest. For this case, we have a stiffness matrix given by

$$\mathcal{C} = \begin{pmatrix} C_{11} & C_{12} & C_{13} & 0 & C_{15} & 0 \\ C_{12} & C_{22} & C_{33} & 0 & C_{25} & 0 \\ C_{13} & C_{23} & C_{33} & 0 & C_{35} & 0 \\ 0 & 0 & 0 & C_{44} & 0 & C_{45} \\ C_{15} & C_{25} & C_{35} & 0 & C_{55} & 0 \\ 0 & 0 & 0 & C_{46} & 0 & C_{66} \end{pmatrix} \qquad (8.19)$$

with simplified Christoffel elements

$$\lambda_{11} = C_{66} \cos^2 \theta_r + C_{55} \sin^2 \theta_r$$

$$\lambda_{22} = C_{22} \cos^2 \theta_r + C_{44} \sin^2 \theta_r$$

$$\lambda_{33} = C_{44} \cos^2 \rho_r + C_{33} \sin^2 \theta_r$$

$$\lambda_{23} = \pm \frac{1}{2}(C_{23} + C_{44}) \sin \theta_r + \cos \theta_r \qquad (8.20)$$

$$\lambda_{13} = C_{46} \cos^2 \theta_r + C_{35} \sin^2 \theta_r$$

$$\lambda_{12} = \pm \frac{1}{2}(C_{25} + C_{46}) \sin \theta_r \cos \theta_r$$

and a governing equation given by

$$\begin{bmatrix} \lambda_{11} - \rho v^2 & \pm \lambda_{12} & \lambda_{13} \\ \pm \lambda_{12} & \lambda_{22} - \rho v^2 & \pm \lambda_{23} \\ \lambda_{13} & \pm \lambda_{23} & \lambda_{33} - \rho v^2 \end{bmatrix} \begin{bmatrix} \alpha_1 \\ \alpha_2 \\ \alpha_3 \end{bmatrix} = \begin{bmatrix} 0 \\ 0 \\ 0 \end{bmatrix} \qquad (8.21)$$

where the "+" sign applies to wave normal l_r incident on the back surface and the "−" sign applies to the wave normal l_{re} reflected from the back surface.

Clearly, the velocities (eigenvalues) are the same, and if $(\alpha_1, \alpha_2, \alpha_3)$ is a solution corresponding to the "+" sign (incident case); $(\alpha_1, -\alpha_2,$

α_3) must be a solution for the refracted wave. Similarly, for the energy flux, we have

$$S_l = \frac{C_{ijklm}\alpha_i\alpha_m l_j}{\rho V}$$

Hence,

$$S_1 = \frac{1}{\rho v}\{\cos\theta_r[C_{66}\alpha_1\alpha_2 + C_{12}\alpha_1\alpha_2 + C_{25}\alpha_2\alpha_3 + C_{46}\alpha_2\alpha_3]$$

$$+ \sin\theta_r[C_{15}\alpha_1^2 + C_{55}\alpha_1\alpha_3 + C_{46}\alpha_2^2 + C_{13}\alpha_1\alpha_3 + C_{35}\alpha_3^2]\}$$

$$S_2 = \pm\frac{1}{\rho v}\{\cos\theta_r[C_{66}\alpha_1^2 + C_{46}\alpha_1\alpha_3 + C_{22}\alpha_2\alpha_3 + C_{46}\alpha_1\alpha_3 + C_{44}\alpha_3^2]$$

$$\quad (8.22)$$

$$+ \sin\theta_r[C_{25}\alpha_1\alpha_2 + C_{46}\alpha_1\alpha_2 + C_{44}\alpha_2\alpha_3 + C_{23}\alpha_2\alpha_3]\}$$

$$S_3 = \frac{1}{\rho v}\{\cos\theta_r[C_{46}\alpha_1\alpha_2 + C_{25}\alpha_1\alpha_2 + C_{23}\alpha_2\alpha_3 + C_{44}\alpha_2\alpha_3]$$

$$+ \sin\theta_r[C_{55}\alpha_1^2 + 2C_{35}\alpha_1\alpha_3 + C_{44}\alpha_2^2 + C_{33}\alpha_j^2]\}$$

or

$$S^+ = \begin{pmatrix} S_1 \\ S_2 \\ S_3 \end{pmatrix} \qquad \text{for the ``+'' case and}$$

$$S^- = \begin{pmatrix} S_1 \\ -S_2 \\ S_3 \end{pmatrix} \qquad \text{for the ``−'' case.}$$

Geometrically, this can be illustrated (here shown for a two-dimensional case) using the illustration shown in Figure 8.5. Since the xz plane is a plane of material symmetry, the top half of the figure must be a mirror image of the bottom half as shown. Using the same nomenclature as

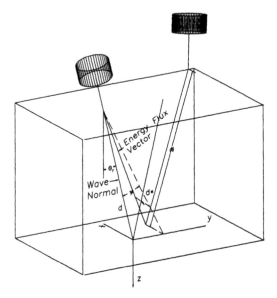

FIGURE 8.5 Experimental geometry for immersion measurements: pulse echo.

for the through transmission case, the pertinent energy flux vectors become

$$S^r = V_{group} \begin{pmatrix} \sin \phi \\ \cos \phi \cos (\theta_r + \psi) \\ \cos \phi \sin (\theta_r + \psi) \end{pmatrix} \tag{8.23}$$

and

$$S^l = V_{group} \begin{pmatrix} \sin \phi \\ -\cos \phi \cos (\theta_r + \psi) \\ \cos \phi \sin (\theta_r + \psi) \end{pmatrix} \tag{8.24}$$

and $S \cdot l = V_{group} \cos \phi \cos \psi = V_{phase}$ for both waves.

The geometry for this experimental configuration is shown in Figure 8.8. The measured time delay between the front surface arrival and back surface arrival will be

$$\Delta t = \frac{2dg}{V_{group}} - \frac{\bar{d} \sin \theta_i}{V_w}$$

$$= \frac{2dg}{V_{group}} - \frac{2dg \cos \phi \sin (\theta_r + \psi) \sin \theta_i}{V_w} \tag{8.25}$$

Using Equation (8.24), this expression can be rewritten as

$$\Delta t = 2 \, dg \cos \phi \left[\frac{\cos \psi}{V_{\text{phase}}} - \frac{\sin (\theta_r + \psi)}{V_w} \sin \theta_i \right] \qquad (8.26)$$

and from Equation (8.8)

$$\Delta t = \frac{2 \, dg}{\cos (\theta_r + \psi)} \left[\frac{\cos \psi}{V_{\text{phase}}} - \frac{\sin (\theta_r + \psi)}{V_w} \sin \theta_i \right]$$

Applying Snell's law, Equation (8.4), yields

$$\Delta t = \frac{2 \, dg}{\cos (\theta_r + \psi)} \left[\frac{\cos \psi}{V_{\text{phase}}} - \frac{\sin (\theta_r + \psi) \sin \theta_r}{V_{\text{phase}}} \right]$$

$$= \frac{2 \, dg}{V_{\text{phase}}} \left[\frac{\cos \psi - \sin (\theta_r + \psi) \sin \theta_r}{\cos (\theta_r + \psi)} \right]$$

Expanding the $\sin (\theta_r + \psi)$ term

$$\Delta t = \frac{2 \, dg}{V_{\text{phase}}} \left[\frac{\cos \psi - \sin^2 \theta_r \cos \psi - \cos \theta_r \sin \theta_r \sin \psi}{\cos (\theta_r + \psi)} \right]$$

Substituting $1 - \cos^2 \theta_r$ for $\sin^2 \theta_r$

$$\Delta t = \frac{2 \, d}{V_{\text{phase}}} \left[\frac{\cos \psi - \cos \psi + \cos^2 \theta_r \cos \psi - \sin \theta_r \cos \theta_r \sin \psi}{\cos (\theta_r + \psi)} \right]$$

$$= \frac{2 \, d}{V_{\text{phase}}} \left[\frac{\cos \theta_r (\cos \theta_r \cos \psi - \sin \theta_r \sin \psi)}{\cos (\theta_r + \psi)} \right]$$

$$= \frac{2 \, d}{V_{\text{phase}}} \cos \theta_r = \frac{2 \, d}{V_{\text{phase}}} \sqrt{1 - \left(\frac{V_{\text{phase}} - \sin \theta_i}{V_w} \right)^2}$$

Solving this expression for the phase velocity yields the final result

$$V_{\text{phase}} = \left[\left(\frac{\Delta t}{2 \, d} \right)^2 + \left(\frac{\sin^2 \theta_i}{V_w} \right)^2 \right]^{-1/2}$$

and, again, the phase velocity can be directly recovered from the time delay measurements.

CHAPTER 9

Ultrasonic Modulus Measurements in Composite Media

One of the principal objectives of nondestructive testing is a quantitative measure of material response. This is best accomplished in anisotropic materials through elastic modulus reconstruction from multiple bulk or guided wave measurements. Then by scanning the transducers over the part it is possible to obtain a mapping of anisotropic material properties on a local basis. It should be pointed out that measurements of this type provide considerably more insight into composite behavior than conventional amplitude-based searches for gross defects, as now material stiffnesses are directly measured, not inferred. In this section, a survey of the many experimental methods developed for elastic modulus reconstruction is presented including bulk and guided wave approaches as well as point/line source techniques.

SECTIONING TECHNIQUES (BULK WAVES)

For thick composites, one direct way of approaching the problem (providing you don't mind destroying part of your sample) is to section the material in such a way as to take optimal advantage of the symmetry of the problem. Much of the early research in this area was, in fact, conducted along these lines [61–68]. For simplicity, let us first consider an example of a unidirectionally reinforced material. As shown in Table 9.1, if one assumes transversely isotropic symmetry, all five elastic constants can be directly measured with a set of simple measurements (or nine if orthotropic symmetry is assumed). One only needs to control the direction of propagation and the mode excited. For these measurements, it is essential that the symmetry directions be known prior to

Table 9.1. Governing Equations for Elastic Modulus Determination using Sectioning Techniques (after Kriz and Stinchcomb [66]).

Specimen No.	Wave Normal Direction Cosines	Particle Direction	Type of Wave	Equation Relating Elastic Moduli and Phase Velocity
1	$\nu_1 = 1$	X_1	L	$C_{11} = \rho c_1^2$
	$\nu_2 = 0$	X_2	T	$C_{66} = \rho c_2^2$
	$\nu_3 = 0$	X_3	T	$C_{55} = \rho c_3^2$
2	$\nu_1 = 0$	X_1	T	$C_{66} = \rho c_1^2$
	$\nu_2 = 1$	X_2	L	$C_{22} = \rho c_2^2$
	$\nu_3 = 0$	X_3	T	$C_{44} = \rho c_3^2$
3	$\nu_1 = 0$	X_1	T	$C_{55} = \rho c_1^2$
	$\nu_2 = 0$	X_2	T	$C_{44} = \rho c_2^2$
	$\nu_3 = 1$	X_3	L	$C_{33} = \rho c_3^2$
4	$\nu_1 = 0$	X_1	T	$C_{66} + C_{55} = 2\rho c_1^2$
	$\nu_2 = 1/\sqrt{2}$	$X_2 X_3$ Plane	L	$C_{23} = \sqrt{(C_{22} + C_{44} - 2\rho c_2^2)(C_{44} + C_{33} - 2\rho c_2^2)} - C_{44}$
	$\nu_3 = 1/\sqrt{2}$	$X_2 X_3$ Plane	T	$C_{23} = \sqrt{(C_{22} + C_{44} - 2\rho c_3^2)(C_{44} + C_{33} - 2\rho c_3^2)} - C_{44}$
5	$\nu_1 = 1/\sqrt{2}$	$X_1 X_3$ Plane	QL	$C_{13} = \sqrt{(C_{11} + C_{55} - 2\rho c_2^2)(C_{55} + C_{33} - \rho 2 c_2^2)} - C_{55}$
	$\nu_2 = 0$	X_2	T	$C_{66} + C_{44} = 2\rho c_1^2$
	$\nu_3 = 1/\sqrt{2}$	$X_1 X_3$ Plane	QT	$C_{13} = \sqrt{(C_{11} + C_{55} - 2\rho c_3^2)(C_{55} + C_{33} - 2\rho c_3^2)} - C_{55}$
6	$\nu_1 = 1/\sqrt{2}$	$X_1 X_2$ Plane	QL	$C_{12} = \sqrt{(C_{11} + C_{55} - 2\rho c_2^2)(C_{66} + C_{22} - 2\rho c_2^2)} - C_{66}$
	$\nu_2 = 1/\sqrt{2}$	$X_1 X_2$ Plane	QT	$C_{12} = \sqrt{(C_{11} + C_{66} - 2\rho 3^2)(C_{66} + C_{22} - 2\rho c_3^2)} - C_{66}$
	$\nu_3 = 0$	X_3	T	$C_{55} + C_{44} = 2\rho c_1^2$

Courtesy Society for Experimental Mechanics, Inc.

sectioning the samples. First, let us see how far we can get by restricting our attention to symmetry axes alone. A cubical sample is cut with one of the sides cut parallel to the fiber axis. Longitudinal velocity measurement along the three axes of the cube immediately yields C_{11}, C_{22}, and C_{33}. Here, we have our first check of the validity of our symmetry. The assumption of transverse isotropy requires that $C_{11} = C_{22}$. Shear wave propagation along the same three axes yields the shear moduli, C_{44}, C_{55} and C_{66}. Here, we again encounter some redundancy in our measurements. For wave propagation along the fiber axis (or X_3 direction), two independent polarizations are possible. Hence, we obtain values for C_{44} and C_{55} for waves whose particle displacement is polarized in the X_2 and X_1 directions, respectively. Similarly, for wave propagation in the X_2 direction (perpendicular to the fibers) shear velocities yield C_{44} and C_{66} and for the X_1 direction C_{55} and C_{66}.

Comparison of measurements of a given modulus provides a good estimate of the measurement uncertainty present in the system. Also, a check of the assumption of transverse isotropy is provided as $C_{44} = C_{55}$ if the assumption is valid. At this point we have been able to determine several, but not all, of the moduli appropriate to our composite model. Yet, these completely exhaust all possible (normal incidence) velocity measurements, for the cube. To proceed further, the only viable alternative is to propagate a wave (or multiple waves) in a nonsymmetry direction and deal with the consequences (impure modes energy flux deviation). Typically, wave propagation at 45° with respect to one of the symmetry axes is chosen. As can be seen in Table 9.1, the algebraic complexity of the equations is increased dramatically if a nonsymmetry direction is chosen. However, since many of the moduli are known (from the previous measurements on the cube) the algebra is straightforward and tractable. As pointed out by Kriz and Stinchcomb [66], care must be taken in the sample design to insure that edge effects are not introduced into the data due to energy flux deviation from the wave normal, as illustrated in Figure 9.1. This will be the case providing

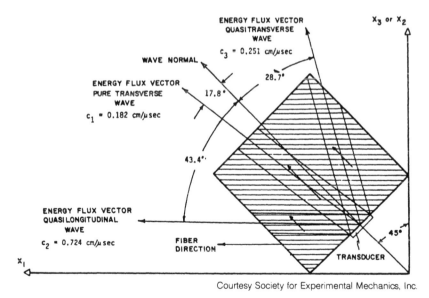

Courtesy Society for Experimental Mechanics, Inc.

FIGURE 9.1 Effect of energy flux deviation from wave normal on specimen design (after Kriz and Stinchcomb [66]).

the width of the sample is sufficiently large to permit the undestructed travel of the wave through the sample, i.e.,

$$\frac{w}{2} \geq \frac{d}{2} + t \tan \phi$$

Using this approach, all pertinent elastic moduli can be determined. This allows us to examine the validity of some of the assumptions made about the symmetry of the problem. In the boron-epoxy samples studied by Tauchert and Guselzu [61], it was found that there were some deviations from the transversely isotropic symmetry usually assumed for this problem. For example, C_{11} and C_{22} as determined ultrasonically were 27.9 and 31.0 $\times$ 10^{10} dynes/cm^2, respectively, although they should be equal under the assumption of transverse isotropy. Similarly, results were obtained for other ostensibly identical moduli and corroborated in later work by Sachse [63] on the same material. These results are also in agreement with results on other (non-composite) anisotropic media with hexagonal symmetry. Part of the explanation for this discrepancy lies in the way in which the specimens were manufactured. As pointed out by Sachse [63], pressure is often preferentially applied to the sample during cure in a single direction (say along the X_1 axis). This produces a texture in the sample due to the flow of the resin in a direction perpendicular to the reinforcing fibers. Thus, orthotropic symmetry may be a better assumption for unidirectionally reinforced composites. However, it should also be pointed out that similar discrepancies were observed in the shear moduli which cannot be explained by processing conditions. A shear wave propagating in the X_1 direction with a particle displacement in the X_2 direction should travel with exactly the same speed as a shear wave propagating in the X_2 direction with a displacement polarized in the X_1 direction. Yet, as shown in Table 9.2, Sachse [63] found that this was not the case. Although these discrepancies are not large, they do bring several of our usual property assumptions into question.

Another point of considerable interest is the degree to which ultrasonic measurements agree with conventional mechanical measurements of the same material properties. The results for the moduli on the diagonal of the stiffness matrix are typically found to be in excellent agreement, as shown in the metal matrix work in Figure 9.2 at least as far as the two moduli shown are concerned. However, as seen in the table of data from Kriz and Stinchcomb [66] in Table 9.3, the degree of agreement depends heavily on the modulus under consideration. For cases where direct measurement is possible, the data are generally

Table 9.2. Elastic Moduli Results as Determined Ultrasonically (after Sachse [63]).

	Equation from Table 1(a) Used to Evaluate	Previous Work [7] (Zero-frequency Limit)	Present Work (10MHz)
C_{11}	(a)	27.9	24.8
C_{22}	(d)	31.0	31.3
C_{33}	(g)	209.	229.
C_{44}	(f, i)	8.48	8.54, 10.0
C_{55}	(c, h)	7.79	7.27, 8.29, 9.30
C_{66}	(b, e)	5.90	6.03, 7.04
C_{23}	(j)	23.1	25.3
C_{13}	(l)	26.1	40.9
C_{12}	(n)	13.7	15.6

in agreement. For moduli that require coupled measurements, typically the off-diagonal terms C_{12}, C_{13}, and C_{23}, the error can be as high as 104% as for the C_{12} measurement here. It should be pointed out that sectioning suffers from several important limitations:

(1) Obviously, it is destructive and unsuitable for quality control except on a limited basis.

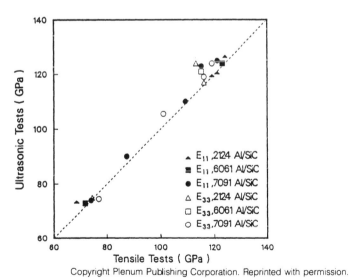

FIGURE 9.2 Comparison of ultrasonic and mechanical test results for elastic moduli (after Jeong et al. [67]).

Table 9.3. Comparison of Mechanical and Ultrasonic Test Results (after Kriz and Stinchcomb [63]).

Specimen No.	Wave Normal Direction Cosine	Displacement Direction	Wave Type	Wave Speed cm/µsec	Experimental Stiffness GPa	Symbol	Predicted Stiffness GPa $V_f = 0.67$	Percent Difference
					AS-3501			
1	$v_1 = 1$	X_1	L	0.974	153	C_{11}	161	5
	$v_2 = 0$	X_2	T	0.224	8.10	C_{66}	7.10	14
	$v_3 = 0$	X_3	T	0.224	8.10	C_{55}	7.10	14
2	$v_1 = 0$	X_1	T	0.230	8.54	C_{66}	7.10	20
	$v_2 = 1$	X_2	L	0.319	16.4	C_{22}	14.5	13
	$v_3 = 0$	X_3	T	0.164	4.31	C_{44}	3.63	19
3	$v_1 = 0$	X_1	T	0.212	7.26	C_{55}	7.10	2
	$v_2 = 0$	X_2	T	0.156	3.90	C_{44}	3.63	7
	$v_3 = 1$	X_3	L	0.311	15.6	C_{33}	14.5	7
4	$v_1 = 0$	X_1	T	0.217	15.2	$C_{66} + C_{55}$	14.2	7
	$v_2 = 1/\sqrt{2}$	$X_2 X_3$ Plane	L	0.308	6.70	C_{23}	7.24	4
	$v_3 = 1/\sqrt{2}$	$X_2 X_3$ Plane	T	0.158	7.91	C_{23}	7.24	9
5	$v_1 = 1/\sqrt{2}$	$X_1 X_3$ Plane	QL	0.680	—	C_{13}	6.50	—
	$v_2 = 0$	X_2	T	0.192	11.9	$C_{66} + C_{44}$	10.7	11
	$v_3 = 1/\sqrt{2}$	$X_1 X_3$ Plane	QT	No Data	—	C_{13}	6.50	—
6	$v_1 = 1/\sqrt{2}$	$X_1 X_2$ Plane	QL	0.679	—	C_{12}	6.50	—
	$v_2 = 1/\sqrt{2}$	$X_1 X_2$ Plane	QT	0.266	13.3	C_{12}	6.50	104
	$v_3 = 0$	X_3	T	0.192	11.8	$C_{55} + C_{44}$	10.7	10

Courtesy Society for Experimental Mechanics, Inc.

114

Table 9.3. (continued).

Wave Speed cm/μsec	Experimental Stiffness GPa	Symbol	Predicted Stiffness GPa $V_f = 0.67$	Percent Difference
		T300/5208		
0.987	154	C_{11}	161	4
0.223	7.84	C_{66}	7.10	10
0.223	7.84	C_{55}	7.10	10
0.223	7.84	C_{66}	7.10	10
0.315	15.7	C_{22}	14.5	8
0.162	4.14	C_{44}	3.63	14
0.223	7.84	C_{55}	7.10	10
0.160	4.05	C_{44}	3.63	12
0.308	15.0	C_{33}	14.5	3
0.223	16.1	$C_{66} + C_{55}$	14.2	13
0.314	7.58	C_{23}	7.24	5
0.161	7.10	C_{23}	7.24	2
0.707	—	C_{13}	6.50	—
0.196	12.2	$C_{66} + C_{44}$	10.7	14
0.260	6.96	C_{13}	6.50	7
0.685	—	C_{12}	6.50	—
0.261	9.09	C_{12}	6.50	40
0.197	12.3	$C_{55} + C_{44}$	10.7	15

115

(2) The technique is less well suited for thin materials than for thick materials, as wave propagation in three orthogonal directions is the preferred means of implementing this technique. As many composite applications call for thin plate structures, this is a serious limitation.

(3) As this is a contact approach, it is best suited for spot inspection. Scanning large areas with contact transducers is feasible, but highly impractical. This is particularly true for shear transducers and the high-viscosity couplants required to introduce shear waves into a specimen. Furthermore, unless sectioning is permitted, one can only measure a limited number of elastic constants with this approach.

Nevertheless, sectioning provides a reliable means of elastic property measurement in anisotropic media. With multiple ray paths it is possible to accurately measure all pertinent moduli, regardless of the symmetry present.

IMMERSION TECHNIQUES (BULK WAVES)

As demonstrated in the previous section, direct contact methods for ultrasonic measurement are accurate and reliable but suffer from important limitations, notably the need for sample sectioning if all composite moduli are to be determined. Immersion techniques offer several major advantages for modulus determination in that they are

- rapid
- nondestructive
- compatible with existing equipment for composite inspection
- suitable for scans of large scale parts

Furthermore, the accuracy of velocity measurements via immersion is comparable to that of measurements via contact, as illustrated in Figure 9.3 for a metal-matrix SiC-Al composite.

As with sectioning techniques, most of the immersion work on composites has been directed at unidirectionally reinforced materials. One of the earliest studies of modulus determination using velocity measurements was performed by Markham [68] on unidirectionally reinforced graphite-epoxy using equipment developed for single crystal studies. The bulk of this work was performed using a through transmission approach as shown in Figure 9.4. In this way, both quasilongitudinal and quasitransverse waves can be generated over a wide range

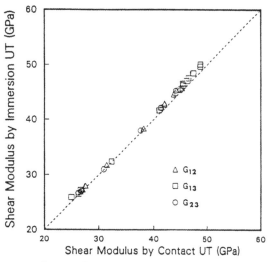

FIGURE 9.3 Comparison of immersion and contact results (after Jeong et al. [67]).

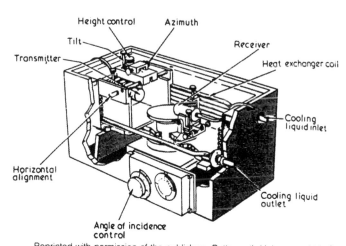

FIGURE 9.4 Ultrasonic goniometer (after Markham [68]).

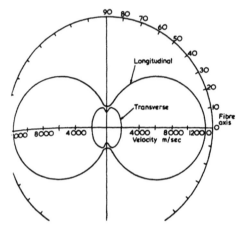

FIGURE 9.5 Velocity surface section—experimental (after Markham [68]).

of inspection angles, at least until the critical angles for each mode is reached. Transit time was measured by comparing the arrival times for signal paths with the sample inserted between the transducers (1) and with sample removed (2). Hence, for normal incidence

$$\frac{1}{V_L} = \frac{1}{V_w} - \frac{\tau}{d}$$

where

V_L = longitudinal wave velocity in the solid
V_w = longitudinal wave velocity in the liquid
τ = arrival time difference

Markham's procedure for immersion velocity measurements relies upon an ultrasonic goniometer (Figure 9.5) for accurate angular measurements. For through transmission measurements, the difference in arrival times for an acoustic path (1) solely in the water and (2) with the sample inserted must be determined. Then,

$$V = \frac{1}{V_w} - \frac{\tau}{d}$$

where

V = phase velocity of sensed wave
τ = measured time delay
d = specimen thickness

It should be pointed out that Markham, like many others, omitted any compensation for energy flux deviation from the wave normal. Fortuitously, as recently proved by Pearson and Murri [69], correction for energy deviation for this differential, through transmission experiment with the direct water path signal as reference, is not necessary, as the compensating terms for the changes in ray path in the specimen (group vs. phase) and water path length precisely offset one another.

Therefore, the above expressions for determination of the phase velocity (regardless of wave mode) is valid, and Markham had a viable means of determining the two possible acoustic velocities in immersion for any direction within the specimen. Typical experimental results are shown in Figure 9.5 where the velocity surface section (in the fiber reinforcement plane) are presented for a carbon fiber-epoxy matrix. The velocity of wave propagation in a non-symmetry direction is usually a non-linear function of several different moduli. Hence, in order to overcome the difficulty of inverting a coupled set of non-linear equation, Markham recommended a multi-step procedure for modulus determination which relies upon direct velocity measurements (as does the sectioning techniques discussed in the previous chapter) for the determination of four of the five moduli. The shear moduli are directly measured using a mode conversion block (here the front surface signal is used as reference) to generate normal incidence shear waves in the sample. By rotating the specimen, both C_{44} and C_{66} ($=C_{11} - C_{12}/2$) can be directly determined.

The longitudinal constants C_{11} (hence C_{12} from above) and C_{33} are determined directly from longitudinal wave propagation along ray path perpendicular and parallel to the fiber axis. As was seen earlier in the discussion of the sectioned samples, the remaining constant C_{13} cannot be directly measured in this fashion. Here, Markham used quasitransverse velocity measurements at an oblique angle of incidence, the known variation of wave speed with direction of propagation and the four moduli already measured for this determination. Markham does bring out an important point regarding the C_{13} measurement. For unidirectional materials, this constant has the "least influence" on acoustic wave propagation. Hence, measurement of this constant is the most difficult and error-prone of all the measurements. As can be seen, Markham's approach is best suited for thick composites, as direct velocity measurements are required. Also, using an additional mode conversion block presents several problems, including compensating for propagation time in the adhesive bond. Clearly, scanning the specimen for global property determination would be difficult with the mode conversion block approach. Nevertheless, Markham's technique represents a nondestructive alternative to sectioning methods with comparable measurement precision.

Like Markham, Smith [70] employed a through transmission arrangement to measure the velocities of propagation for multiple angles of oblique incidence (Figure 9.6). Here, however, a somewhat different approach to modulus determination was employed. The normal stiffnesses C_{11} and C_{33} were directly determined from normal incidence longitudinal wave speed measurements in directions perpendicular and parallel to the reinforcing fibers. The remaining moduli were determined by measuring the angular dependence of the quasitransverse mode of propagation, i.e., experimental mapping out of what is essentially a velocity surface section for a particular mode. C_{44} and C_{13} are determined by adjusting their values to best fit (in the sense of least square error) the known angular dependence of the velocity of the mode (1-3 plane). The remaining modulus, C_{66}, is determined in a similar fashion by turning the specimen 90° and mapping out the transverse wave speed for the 1-2 plane. Figure 9.7 shows that good agreement obtained by Smith for the experimentally measured and predicted values of the effective shear modulus (ρ/V^2_{QT}) in the direction ϕ with respect to the fiber axis, from the Christoffel equations discussed earlier.

Gieske and Allred [71] used a combination of contact and immersion measurements to determine all of the components of a Boron-Aluminum composite assuming orthotropic symmetry. They found the assumption of transverse isotropy to be 15%–25% in error, much as the other workers have found [61, 63].

Hosten and co-workers [72–75] have extensively studied the characterization of composites with both anisotropic and viscoelastic properties through measurement of both the velocity and/or attenuation along multiple ray paths. This group has utilized an approach that is based upon the use of an ultrasonic goniometer to measure the angular dependence of ultrasonic velocity along with an iterative curve-fitting algorithm (based upon Newton's method) to extract the C_{ij}'s which yield the best fit to the data. Typical results are shown in Figure 9.8. They have also used this approach to good advantage in tracking the progress of densification in carbon–carbon composites. This work illustrates the directional property sensitivity, which is one of the major advantages of this technique. Carbon–carbon composites consist of carbon fibers in a carbon matrix and are processed in a multistep process. Initially, a phenolic resin/carbon fiber composite is produced. The material is then heated to an elevated temperature to carbonize the matrix. This creates a porous structure, which must be densified to be useful structurally. This is achieved via repeated cycles of chemical

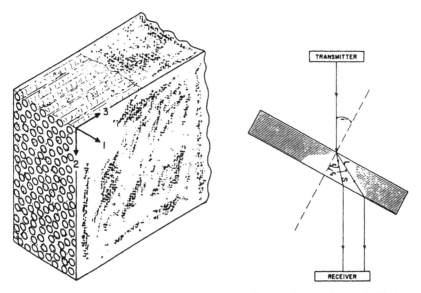

Courtesy American Institute for Physics.

FIGURE 9.6 Geometry for ultrasonic through transmission studies (after Smith [70]).

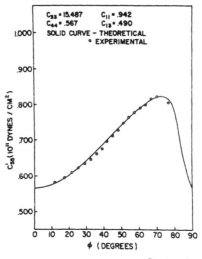

Courtesy American Institute for Physics.

FIGURE 9.7 Variation of shear modulus with orientation: theory vs. experiment (after Smith [70]).

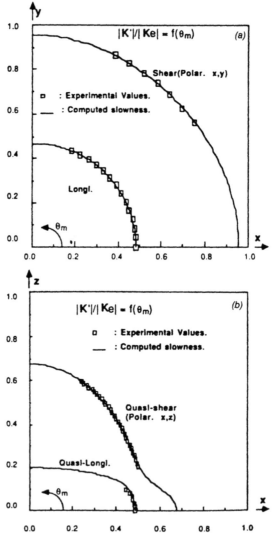

FIGURE 9.8 Slowness curves: experiment and numerical fit (after Hosten et al. [75]).

Table 9.4. Effects of Processing on Carbon–Carbon Composite Moduli
(after Hosten and Tittmann [74]).

Sample	Density	Velocity in the 1 Dir.		C11 C22 C60 C12				Anisotropy Factor
		Longi. mss.	Shear mss.	Mega Phase				$\dfrac{2\ C60}{(C22\text{-}C12)}$
KA-1 Initial State	1.42	3298	1940 contact 1926	15	20	5.3	5.4	0.73
KA-2 First Carbonization	1.32	1523	1520 contact 1505	3.4	13	3.0	1.1	0.52
KA-3 Re-filled	1.47	2687	1890 contact 1870	11	19	5.0	2.4	0.62
KA-4 Second Carbonization	1.38	2190	1740 contact 1728	6.6	24	2.0	2.5	0.40

vapor deposition (CVD) and pyrolysis until a suitably dense microstructure is achieved. Results from the ultrasonic moduli determination
at various stages in the process are presented in Table 9.4. Note the
large drop in the inplane stiffnesses during the initial carbonization.
This is principally due to the thermal expansion mismatch between the
fibers and matrix, which results in extensive local microcracking as
shown in Figure 9.9. Here, the out-of-plane stiffnesses are relatively
unaffected, as expected from the orientation of the microcracks. This
research group has also been the first to characterize both the real and
imaginary parts of the anisotropic stiffness matrix for a lossy composite.
Here, the constitutive relationship for harmonic loading is a stiffness
matrix given by elements of the form

$$C_{ij} = C_{ij} + i\omega\mu_{ij}$$

Here the angular dependence of both velocity and attenuation are
measured to produce slowness surface sections, as usual, and attenuation surfaces as well (Figures 9.8 and 9.11). The introduction of a complex valued stiffness (assuming real values of the frequency) requires a
complex valued wave number as, for example, in the simplest case

$$V_L = \frac{\omega}{K^*} = \sqrt{\frac{C_{11}^*}{\rho}}$$

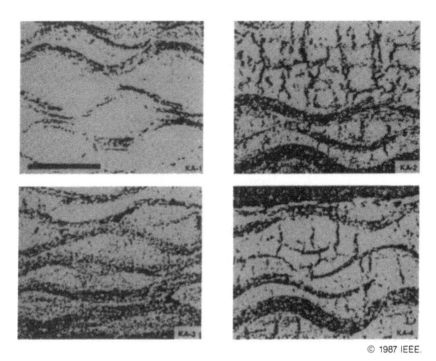

FIGURE 9.9 Matrix microcracking in carbon–carbon composites (after Hosten and Tittmann [74]).

Therefore, $K^* = \beta + i\alpha$ and for, say, harmonic wave propagation in the x direction

$$\mathbf{u}(x, t) = A_0 \boldsymbol{\alpha} e^{-\alpha x} e^{i(\beta x - \omega t)}$$

Although the algebra is somewhat more complicated, similar expressions hold for other modes of propagation and for wave propagation in an arbitrary direction. Typical results for the real and imaginary components of the stiffness matrix for unidirectional graphite-epoxy are presented in Table 9.5.

Table 9.5. Real and Imaginary Parts of Stiffness Matrix:
Unidirectionally Reinforced Graphite–Epoxy (after Hosten et al. [75]).

$$C_{ij} = \begin{bmatrix} 15 & 7.7 & 3.4 & & & \\ 7.7 & 16 & 3.4 & & & \\ 3.4 & 3.4 & 87 & & & \\ & & & 7.8 & & \\ & & & & 7.8 & \\ & & & & & 3.9 \end{bmatrix} \text{GPa.} \qquad \mu_{ij} = \begin{bmatrix} 14 & 6.4 & 46 & & & \\ 6.4 & 11 & 46 & & & \\ 46 & 46 & 513 & & & \\ & & & 4.2 & & \\ & & & & 4.2 & \\ & & & & & 3.4 \end{bmatrix} \text{PaS.}$$

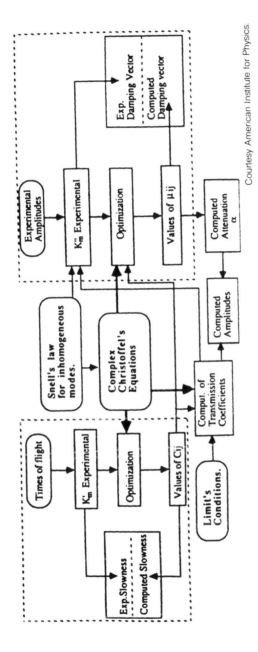

FIGURE 9.10 Schematic diagram for property reconstruction (anisotropic, viscoelastic media) (after Hosten et al. [75]).

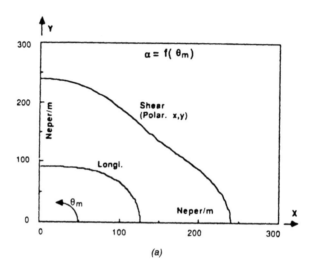

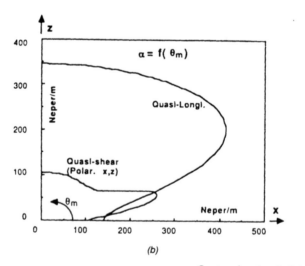

FIGURE 9.11 Attenuation surface sections—unidirectionally reinforced graphite-epoxy (Note: fibers aligned with z axis) (after Hosten et al. [75]).

Most of the work discussed to this point has been directed towards modulus determination at a single point. However, in practice, composite properties are highly sensitive to processing conditions and, as a result, they can be highly nonhomogeneous. Composite manufacture is a complicated process involving large amounts of resin flow (to eliminate trapped gases as well as excess resin) or crystallization (thermoplastics). As it is difficult to maintain a uniform temperature and pressure distribution throughout large-scale composite parts, nonuniform properties result. Anisotropic modulus determination is an ideal way to characterize this behavior.

Kline and co-workers [76–79] have developed both pitch-catch and through transmission techniques suitable for the scans of large-scale, laminated composite parts (not just unidirectional). This approach uses precisely controlled transducer inspection angles to generate the multiple ray paths necessary for complete modulus characterization. The transducer assembly is then scanned over the part and the velocity is determined at each point using either a correlation technique or a relative measurement of velocity changes using a pulse phase locked loop approach as described in Chapter 6. Control of the transducer inspection angles can be totally automated using a maximum transmission criterion to establish the local normal to the specimen surface (in through transmission) or a maximum reflection criterion for each transducer or pulse-echo testing. An RVDT assembly transducer fixture is then used for accurate angular positioning (Figure 9.12). All necessary individual transducer or transducer fixture rotations can be automated using computer-controlled servo motors (Figure 9.13). One may choose to perform all required measurements at each sample position through computer-controlled adjustment of transducer orientation or, if a suitable reference is chosen, to fix inspection angle and perform multiple part scans. It should be pointed out that this is not truly a "point" measurement but rather a volumetric average due to the finite size of the transducer as well as refraction/reflection effects at oblique incidence. In this way, two-dimensional maps of local material behavior may be obtained. Typical results are presented in Figure 9.14 for the stiffness perpendicular to (C_{11}) and parallel to the fiber direction (C_{33}) of a $12'' \times 18''$ unidirectionally reinforced graphite-epoxy panel.

One of the major conclusions to be drawn from this approach is that contour plots of velocities provide new insights into the microstructure of fabricated graphite-epoxy panels. For unidirectionally reinforced panels, regions of constant velocity are often oriented parallel to the reinforcing fibers, and indicate fiber segregation during processing. This is not surprising in light of the fact that the reinforcing fibers restrain

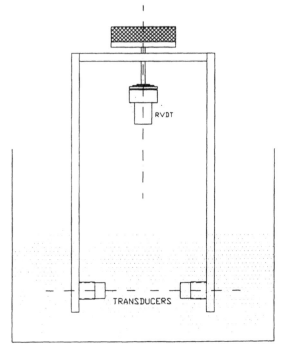

FIGURE 9.12 RVDT assembly for precise control of angular position-through transmission.

FIGURE 9.13 Stepper motor assembly for angular position control-pulse-echo.

128

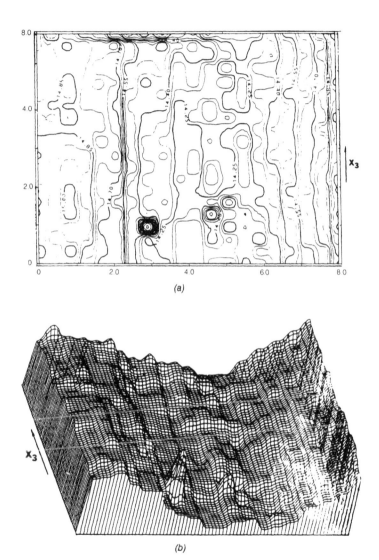

(a)

(b)

FIGURE 9.14 Plot of modulus vs. position in unidirectional *Gr/Ep* sample (after Kline and Chen [76]).

motion in the reinforcing direction but provide little resistance to flow perpendicular to this direction. Previous studies indicate fiber-volume fraction variations of nearly 20% possible for panels of this size. One also can observe anomalies such as high/low modulus at plate edges.

Processing is found to play a key role in determining the local microstructure, particularly for unidirectionally reinforced materials. Note that the local modulus is significantly higher along two of the edges of the panel than it is in the center. This is principally attributable to resin flow during processing. Resistance to flow is relatively low on these edges. This is in contrast to the other two edges, where flow is restricted by the oriented fibers, or in the center of the part. Removal of excess resin and fiber accumulation allows a higher fiber volume function, hence higher modulus can be achieved in these regions. It should be noted that local defects (as would be imaged with a conventional C-scan) also appear prominently in these plots, usually as peaks (or valleys), as illustrated in Figure 9.14. In this case, the indication corresponds to a debond near the bottom surface. Typical echoes (normal incidence) for defects such as these as well as normal regions are shown in Figure 9.15. Results of the five moduli for a given plate were in

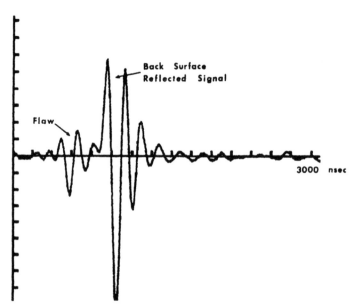

FIGURE 9.15 Normal BSR along with flaw reflection in unidirectional *Gr/Ep* (after Kline and Chen [76]).

substantive agreement both qualitatively and quantitatively with one another. Typical modulus variations from point to point in these panels ranged from 14 to 17%. Defect locations were usually found to be in register with one another from plot to plot.

It should also be pointed out that most work in this area has been performed on unidirectionally reinforced composites. Unfortunately, while this provides a convenient test geometry, it also avoids many of the complexities associated with more practical composite layups. However, applications to both unidirectionally reinforced composites and laminate composites with multiple ply orientation are possible.

Laminated composites, with multiple reinforcement directions, present a considerably more difficult problem. Ideally, one would like to be able to isolate successive echoes from each individual ply in the composite, but this approach is impractical for all but the simplest laminate geometries. Alternatively, if the wavelength of the acoustic waves is large in comparison to the ply thickness, one can treat the laminate as a continuum much in the same manner as fiber and matrix effects are commonly ignored in modeling wave propagation in an individual ply. While a matter of some concern for many graphite-epoxy composites with ply thickness on the order of 0.01 in., this assumption may be overly restrictive, particularly in the low frequency test regime ($\omega \leq 1$ MHz). Typical results for one of the effective inplane stiffnesses C_{33} for a cross ply sample are presented in Figure 9.16. Note that the extent of local property variation is significantly reduced from the unidirectional case due to the restraining effects on the flow of the multiple fiber orientation.

Alternatively, we can assume that the composite is composed of identical transversely isotropic layers whose orientations are known. Since the orientations of each ply may be different, in order to determine the contribution stiffness for each ply in the laminate we must first transform the stiffness components from each ply to a common coordinate system in accordance with the tensor transformation laws for a fourth-order tensor. For this case, a ply whose fibers were oriented at an angle θ w.r.t., the z axis would have a stiffness matrix C_{ij} given by

$$C'_{11} = C_{11} \cos^4 \theta + 2C_{13} \cos^2 \theta \sin^2 \theta + C_{33} \sin^4 \theta$$

$$+ 4C_{55} \cos^2 \theta \sin^2 \theta$$

$$C'_{12} = C_{12} \cos^2 \theta + C_{23} \sin^2 \theta$$

$$C'_{13} = C_{11} \cos^3 \theta \sin \theta + C_{13}[\cos^3 \theta \sin \theta - \cos \theta \sin^3 \theta]$$
$$+ C_{13} \cos^2 \theta \sin^2 \theta - 4C_{55} \cos^2 \theta \sin^2 \theta$$

$$C'_{15} = -C_{11} \cos^3 \theta \sin \theta + C_{13}[\cos^3 \theta \sin \theta - \cos \theta \sin^3 \theta]$$
$$+ C_{33} \cos \theta \sin^3 \theta + 2C_{55}[\cos^3 \theta \sin \theta - \cos \theta \sin^3 \theta]$$

$$C'_{22} = C_{11}$$

$$C'_{23} = C_{12} \sin^2 \theta + C_{23} \cos^2 \theta$$

$$C'_{25} = -C_{12} \cos \theta \sin \theta + C_{23} \cos \theta \sin \theta$$

$$C'_{33} = C_{11} \cos \theta \sin^3 \theta + C_{13}[\cos \theta \sin^3 \theta - \cos^3 \theta \sin \theta]$$
$$+ 4C_{55} \cos^2 \theta \sin^2 \theta$$

$$C'_{35} = -C_{11} \cos \theta \sin^3 \theta + C_{13}[\cos \theta \sin^3 \theta - \cos^3 \theta \sin \theta]$$
$$+ C_{33} \cos^3 \theta \sin \theta + 2C_{55}[\cos \theta \sin^3 \theta - \cos^3 \theta \sin \theta]$$

$$C'_{44} = C_{44} \cos^2 \theta + C_{66} \sin^2 \theta$$

$$C'_{46} = C_{44} \cos \theta \sin \theta + C_{66} \cos \theta \sin \theta$$

$$C'_{55} = C_{11} \cos^2 \theta \sin^2 \theta + C_{33} \cos^2 \theta \sin^2 \theta - 2C_{13} \cos^2 \theta \sin^2 \theta$$
$$+ C_{55} (\cos^2 \theta - \sin^2 \theta)^2$$

$$C'_{66} = C_{66} \cos^2 \theta + C_{44} \sin^2 \theta$$

$$C'_{14} = C'_{16} = C'_{24} = C'_{26} = C'_{34} = C'_{36} = C'_{45} = C'_{56} = 0$$

Then we can treat the problem as having five unknowns, the five elastic moduli appropriate for a transversely isotropic ply, or nine unknowns for a woven ply laminate. Due to extensive mode conversion at each

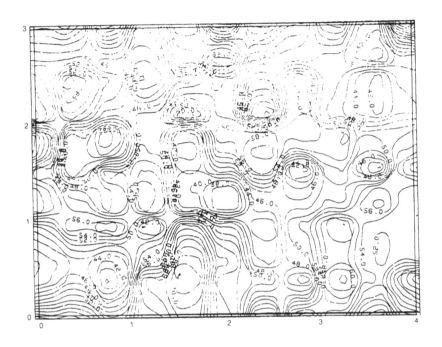

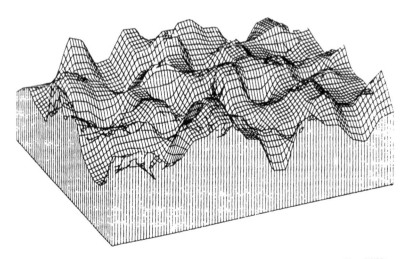

FIGURE 9.16 Mechanical property map (C_{33}) for cross ply sample (after Kline [77]).

133

ply interface, it is desirable to concentrate on only the fastest arriving wave (quasilongitudinal) in the laminate to avoid ambiguity in mode identification. The major drawbacks with this approach are that ply to ply variations through the thickness are ignored (obviously a questionable assumption in very thick laminates) and any ply misorientation from the assumed layup sequence will adversely influence the accuracy of the results. Furthermore, since only the initial QL arrivals are of practical use, sensitivity to certain moduli may be severely restricted.

A similar approach was used to investigate the behavior of a full-scale composite component, the space shuttle brake shoe. This is a woven (orthotropic) ring structure 18″ OD, 12″ ID, and 0.857″ thick. The thickness of the composite necessitated the use of relatively low frequency transducers (500 kHz center frequency). An interesting aspect of this investigation was the use of a complementary NDE technique, radiography, to determine local density in the porous carbon–carbon sample. This is needed as the density ρ figures prominently in the Christoffel equation

$$\det \left[\underset{\sim}{\lambda} - \rho v^2 \underset{\sim}{I} \right] = 0$$

and any changes in local density would influence the C_{ij} calculations. Local densities, as determined from the radiographs, are used as data input at each measurement point along with the experimentally determined time delays to accurately correct for this phenomenon. A schematic of the experimental setup is shown in Figure 9.17. Also of note is the use of a pulse phase locked loop (P^2L^2) system to track local changes in transit time. This approach, being a relative measuring, requires an accurate initial velocity measurement at a reference point. To do this, a cube of similar material was placed next to the test component during the scan. This cube was subsequently sectioned and, via the techniques described earlier, the elastic moduli determined. These values were used as the basis for calculation of the transit time for each mode of interest at the reference point. Typical results are presented in Figure 9.18. Note the existence of an anomalous region near the 0° position of the scan. This region also appeared to be abnormal in the radiograph, indicating a possible region of incomplete densification.

Hsu and Margetan [80] have developed a unique variation of the conventional (pulse-echo or through transmission) approach to ultrasonic velocity measurements using a modified version of the acousto-ultrasonic technique. The experimental geometry is shown in Figure 9.19. Here, the interest is in the reflected echo pattern, including mode conversion. This approach takes advantage of the fact that, while most

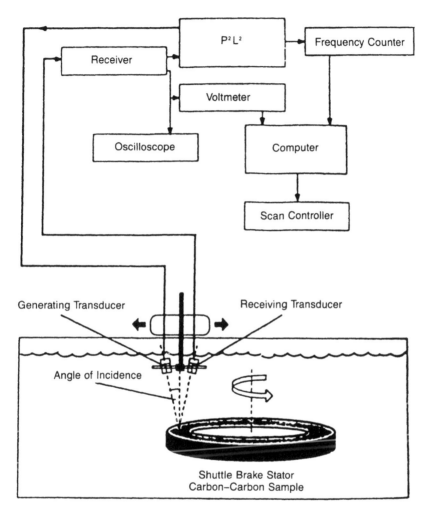

FIGURE 9.17 Experimental schematic-P^2L^2 scan (after Kline et al. [78]).

of the energy in a contact transducer is radiated in the direction normal to transducer surface, there will be some useful energy radiated in all directions in the sample. By analyzing the various echoes in the acoustic signal sensed by the receiving transducer, the velocities of multiple modes of propagation can be determined from a single experiment. The changes in ultrasonic beam orientation which are required for wave surface reconstruction and modulus determination are produced by altering the position of the receiving transducer in a systematic fashion.

C22 MODULUS (in MPa)
(2 IS THE THETA DIRECTION)

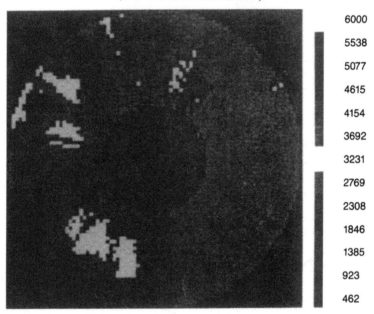

6000
5538
5077
4615
4154
3692
3231
2769
2308
1846
1385
923
462

FIGURE 9.18 Mechanical property map carbon-carbon stator (after Kline et al. [78]).

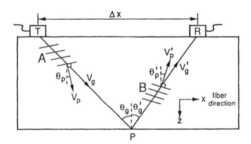

FIGURE 9.19 Experimental geometry-acousto-ultrasonic technique (after Hsu and Margetan [80]).

Typical results comparing theory and experiment for the various echoes identified in a graphite-epoxy sample are shown in Figure 9.20. Note the ability of the technique to accurately map the cusp region of the group velocity curve.

GUIDED WAVE TECHNIQUES

Since the capability to characterize the reflection and transmission spectra for plate wave propagation in composites (immersion testing) is well established (see Chapter 5), several investigators have successfully employed this approach to recover the elastic properties of test panels. Rokhlin and Chimenti [81] have explored the use of an iterative method to achieve this goal. They begin with a set of initial estimates for the elastic moduli. Then, for the known angle of incidence, they find approximate values for the parameter *fh* (frequency thickness) which yield minima in the reflection coefficient spectra from the guessed C_{ij}'s. Using an iterative scheme, the precise *fh* corresponding to minima are identified. After ordering the minima, a measure of the deviation be-

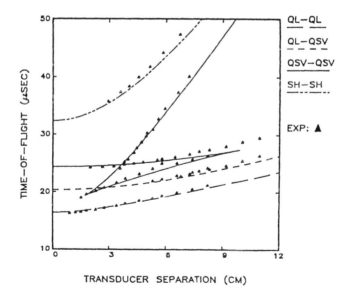

FIGURE 9.20 Comparison of theory vs. experiment for time of flight measurements–acousto-ultrasonic experiment (after Hsu and Margetan [80]).

tween the experimentally identified minima and the guessed values is calculated as

$$\Delta = -\sum [(fh)_i^{calc} - (fh)_i^{meas}]$$

where the sum is over all of the points of the experimentally measured reflection coefficient spectrum. From this function, gradients are calculated and new estimates of the elastic moduli chosen using the Newton–Raphson procedure. The procedure is then repeated until the solution converges to the final answer.

As with all of the available experimental methods, the question of sensitivity is critical. Rokhlin and Chimenti have shown that for certain experimental configurations, particular moduli may have little effect on the experimental results, hence it may be difficult to determine these parameters with any degree of accuracy for these situations. This is illustrated in Figure 9.21 for the two inplane normal stiffness C_{11} and C_{33} (here, 1 is the fiber direction). For incidence angles below the first critical angle, the results depend strongly upon changes in C_{11} but are relatively unaffected by changes in C_{33}. Hence, care must go into the choice of experimental measurements if one wishes to insure meaningful results. The authors had low confidence in results for C_{12} and C_{66} due to the lack of sensitivity to these parameters. Dayal and Kinra [82] have also developed a similar algorithm for elastic moduli reconstruction from plate wave propagation measurements.

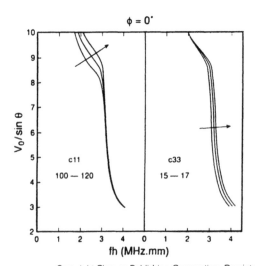

FIGURE 9.21 Sensitivity of guided wave technique to in-plane stiffnesses (after Rokhlin and Chimenti [82]).

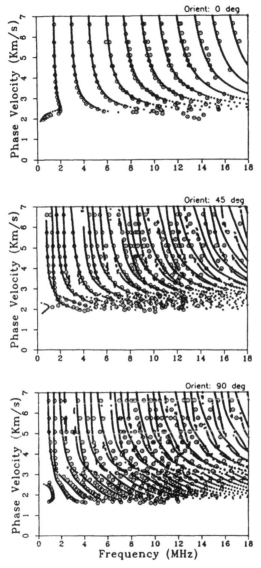

Courtesy American Institute for Physics.

FIGURE 9.22 Dispersion curves using data reconstructed from moduli determined with simplex algorithm (after Karim et al. [84]).

Bar-Cohen, Mal, and co-workers [83–84] have also been very active in this area, with several studies of the problem from both a theoretical and experimental standpoint. They have recently developed a somewhat different inversion scheme based upon a simplex method for the recovery of the elastic moduli from the dispersion curves. Figure 9.22 shows typical results from the simplex algorithm for dispersion curves at 0°, 45°, and 90° w.r.t. reinforcing fibers in a composite plate sample for the reconstructed curves.

While the bulk of work in this area has concentrated on reflection measurements, additional information is present in the transmission spectra. Rokhlin and co-workers [85–86] have used transmission measurements to achieve the same end using the Nayfeh and Chimenti formulation for transmission. Experimental geometry is shown in Figure 9.23 for a novel double transmission/reflection goniometer used in these studies. Figure 9.24 illustrates (a) the normalized transmission coefficient as a function of angle of incidence for a unidirectionally reinforced plate for three different transducer orientations (with respect to the fiber axis) and (b) a comparison of theory and experiment for

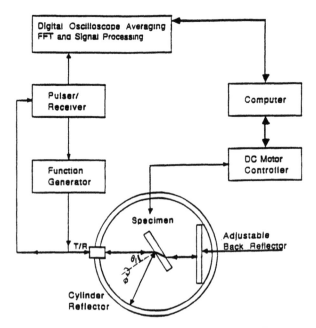

FIGURE 9.23 Experimental setup–double transmission/reflection system (after Rokhlin and Wang [86]).

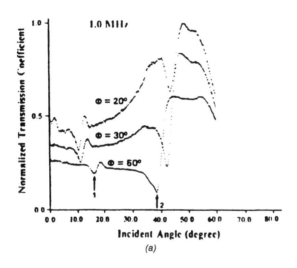

(a)

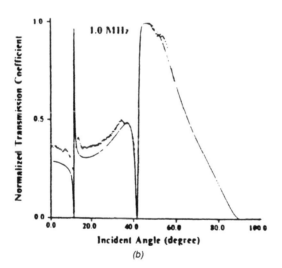

(b)

FIGURE 9.24 Plate mode transmission coefficient vs. incidence angle unidirectional $Gr - Ep$ (after Rokhlin et al. [85]).

141

a typical curve. In order to take optimal advantage of the sensitivity of the measurements to particular elastic moduli, a computer simulation was performed to identify the features of the curves which would be particularly useful in this regard. In this investigation, it was found that the inplane normal stiffnesses were optimally determined from measurements parallel and perpendicular to the axis of fiber reinforcement. The remaining moduli were recovered from measurements in nonsymmetry directions. A least square minimization scheme was used as the basis for the recovery procedure. Using this method, a good match of the theoretical prediction to the experimental data was obtained. Wang and Rokhlin [86] have further advanced their approach to elastic moduli reconstruction using plate modes in the same double transmission/reflection system. Typical results for zero transmission angle as a function of fiber orientation, as predicted from theory as well as with experimental data from the reconstructed elastic constants, are shown in Figure 9.25. Good agreement was observed between these results and direct mechanical property measurements of previous authors, as shown in Table 9.6 for a metal matrix SiC-Al composite.

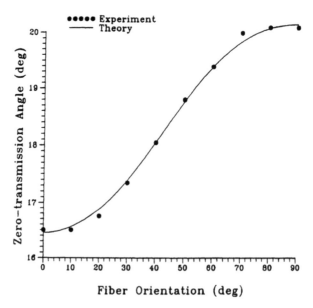

FIGURE 9.25 Plot of zero transmission angle vs. specimen orientation: theory vs. experiment after Rokhlin and Wang [86]).

Table 9.6. Comparison of Plate Wave Moduli with Mechanical Test Results (after Rokhlin and Wang [86]).

	C_{11}	C_{22}	C_{33}	C_{66}	C_{12}	C_{23}	C_{13}
Experiment	121	108	76.6	20.9	65.0	69.6	61.5
Nielsen [6]	125	115	115	28.4	58.1	58.8	58.1
Christensen [7]	123	115	115	28.3	56.0	58.8	56.0

POINT/LINE SOURCE TECHNIQUES

While the majority of experimental work on composites has employed (piezoelectric) plane wave sources, other source/sensor configurations are possible and can be used to perform the same type of measurements. The development of pulsed laser sources with sufficient power to generate elastic waves (primarily via thermal expansion) as well as point detectors (laser interferometer, NBS conical transducer, capacitance transducer, etc.) has been particularly useful in developing these new test methods. Theoretically, the displacement field generated by a point (or extended source) can be calculated (in principle) as the convolution of the source function (**f**) with the dynamic Green's tensor for the material (**G**) as

$$u_i(\mathbf{r}, t) = \int_{V_o} \int_0^T f_j(\mathbf{r}', t) G_{ij}(\mathbf{r}/\mathbf{r}', t \cdot \tau) d\tau dV'$$

The determination of the dynamic Green's function for various problems in anisotropic media remains a subject of intense research interest, and a full treatment of the problem is beyond the scope of this monograph. Payton [87] has published an excellent discussion of the subject for problems of transversely isotropic symmetry, with particular emphasis on an analytical line source calculations. The results are very similar to results from isotropic media, with wave arrivals from the bulk longitudinal and shear waves as well as a sharp discontinuity due to the Rayleigh wave arrival. For propagation in nonsymmetry directions, three bulk wave arrivals (one quasilongitudinal and two quasi-transverse) can be discerned. More recently, Rose and co-workers [88] have developed a numerical code to achieve the same end.

Doyle and Scala [90] have developed a method for determining elastic moduli using a laser line source to generate the acoustic waves. One of the advantages of using a line source is that there is a well-defined wave normal (perpendicular to the source axis); hence both group velocity and the angle between the energy flux vector and the wave normal can be directly measured. Doyle and Scala propose a technique for elastic property reconstruction in materials with transversely isotropic or orthotropic symmetry. The proposed method relies principally on measurements of velocity and energy flux deviation angle for the longitudinal/quasilongitudinal wave. For fiber reinforced composites with transversely isotropic symmetry, the in-plane normal stiffnesses C_{11} and C_{33} are directly measured from the velocities in the 1 and 3 directions. As these are symmetry axes, the energy flux vector and wave normal coincide so there is no discrepancy between phase and group velocity. The remaining two in-plane moduli, C_{13} and C_{55}, are determined from multiple velocity and energy flux deviation angle measurements and a least-squares fit to the data. Doyle and Scala have investigated the potential error associated with a limited number of samples in reconstituting these two stiffness values and recommend that both velocity and deviation angle measurements be made for at least eight different directions to insure the accuracy of the numerical result. The remaining constant (C_{12}) can be measured from the transit time for the through thickness (bulk) shear wave generated with the axis of the line source parallel to the fibers using the relation

$$\rho[V_1^2]^2 = C_{66} = (C_{11} - C_{12})/2$$

as C_{11} is already determined. Alternative procedures using Rayleigh wave speed measurements are also discussed. These results can be extended to materials with orthotropic symmetry using a similar procedure. However, measurement of the arrival times for several ray paths for bulk waves that reflect off the back surface of the composite are required when the assumption of transverse isotropy is no longer valid (Figure 9.26).

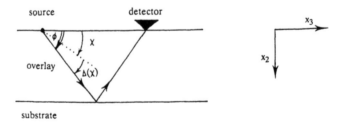

FIGURE 9.26 Thermoelastic line source ray paths (after Doyle and Scala [90]).

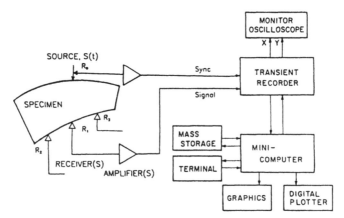

FIGURE 9.27 Point source geometry (after Sachse et al. [91]).

Sachse and co-workers [91, 92] have also employed a similar approach to the problem of elastic constant reconstruction, here with a point rather than line source. The fracture of glass capillary tubes was used to generate the acoustic waves, since in several previous studies this was found to be a good way to experimentally simulate a localized point source varying in time as a Heaviside step function. The experimental setup is illustrated in Figure 9.27. In this arrangement, a single source is used in conjunction with an array of small aperture (1.3 mm) transducers to yield the quasilongitudinal and quasitransverse arrival times for multiple ray paths in the specimen. The Newton–Raphson algorithm is used to solve the resulting coupled set of nonlinear equations for elastic modulus recovery. Figure 9.28 shows typical slowness results for an orthotropic media. The data points are computed values for the slowness where the source and receiver positions were randomly perturbed by 0.1%. The solid lines show the results obtained using the Newton–Raphson algorithm with good agreement found between two approaches. More recently [93], results have been obtained using all three sheets of the slowness surface section for modulus reconstruction (Figure 9.29). Again, good results have been achieved with the data obtained using experimentally determined constants.

TECHNIQUE BASED CRITICAL ANGLE PHENOMENON

Rokhlin and Wang [94] attacked the problem from a somewhat different perspective. Critical angle measurements have been used for

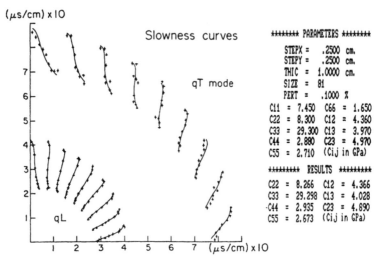

FIGURE 9.28 Reconstructed slowness curves: theory vs. experimental (point source, after Sachse et al. [91]).

quite some time for velocity measurements in isotropic solids. This follows directly from Snell's law, as the maximum value for the sine of the angle of refraction is one. Therefore, for incidence at the critical angle $[\theta_{in}]_{cr}$

$$V_{re} = \frac{V_{in}}{\sin{(\theta_{in})_{cr}}}$$

Similar considerations also hold for anisotropic media. However, as pointed out by Henneke and co-workers [95], energy considerations may supersede Snell's law. There are potential situations where the physically impossible situation of energy being transmitted into the incident media before the critical angle (as given by Snell's law) is reached. Henneke and Jones [96] give one such example for a quartz-water interface and recommend that critical angle be determined from energy flux (not wave normal) being parallel to the media interface. Therefore, the shape of the slowness surface at media boundaries is a critical consideration in interpreting experimental results. Fortunately, for fiber reinforced composites, the symmetry of the problem (here the fiber reinforcement plane is a plane of material symmetry) insures that there is no difference in critical angle as determined from either group

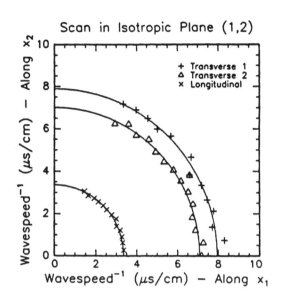

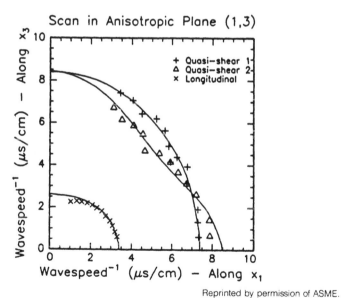

FIGURE 9.29 Slowness surface section showing all three sheets in unidirectional fiberglass polyester (after Sachse et al. [91]).

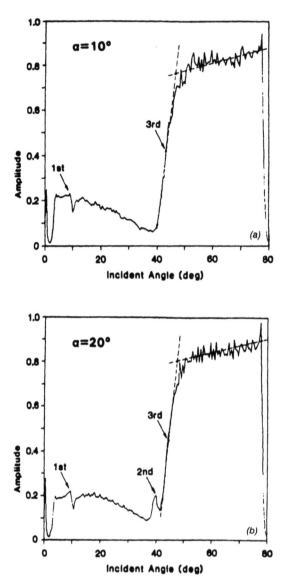

Courtesy American Institute for Physics.

FIGURE 9.30 Amplitude vs. angle of incidence measurements for critical angle determination (after Rokhlin and Wang [94]).

148

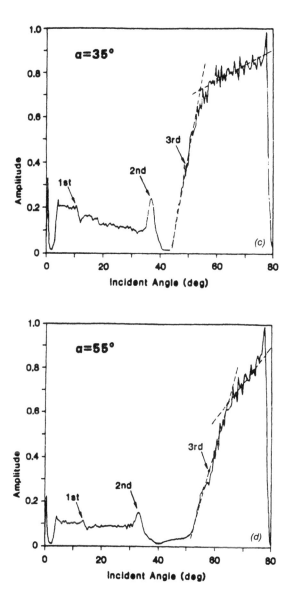

FIGURE 9.30 (continued) Amplitude vs. angle of incidence measurements for critical angle determination (after Rokhlin and Wang [94]).

149

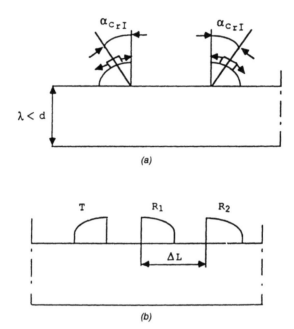

(a)

(b)

FIGURE 9.31 Experimental procedure for subsurface longitudinal wave generation (after Pilarski and Rose [97]).

(energy fllux) or phase (Snell's law) considerations. If one calculates the components of S as a function of refraction angle, as done by Rokhlin and Wang, it can be shown that, at the critical angle, both the energy flux vector (S) and the wave normal (l) will lie in the plane of fiber reinforcement and therefore result in the same limitation on angle of incidence.

This observation is the basis for their experimental technique. In their work, Rokhlin and Wang use an ultrasonic goniometer assembly as described earlier in this chapter (Figure 9.23) to rotate the transducer through various angles of incidence and a peak detector to measure signal amplitude. Here, a double reflection technique was employed so that only a single transducer was required. Figure 9.30 illustrates experimental results for a sample oriented at four different angles with respect to the fiber axis. Here, signal maxima associated with the quasi-longitudinal and two quasitransverse waves can be determined. From these angles, the velocity can be calculated. One then repeats these measurements for multiple angles and uses a curve-fitting technique (here minimization of the sum of squares) to reconstruct the elastic constants in a manner similar to that used by Hosten [72] and others.

In a related vein, Pilarski and Rose [97] studied the waves generated within a sample at the first critical angle as a means for composite modulus determination. This technique was explored because it offers the advantage of being a single-sided measurement. Pilarski and Rose term the longitudinal (or quasilongitudinal) wave generated within a sample at the first critical angle, a sub-surface longitudinal or SSL wave. The experimental procedure for their measurements is illustrated in Figure 9.31. The first critical angle is measured by adjusting the incidence angles for the generating and receiving transducers and identifying the angle associated with maximum received. The time delay is then measured. Then, the receiving transducer is translated a fixed distance and a new time delay measured to provide a means of velocity determination. In this approach to elastic constant measurement, multiple ray types are used. Contact measurements of the longitudinal and two pure shear modes (for wave propagation normal to the plate) were used to measure the three available moduli (C_{22} and C_{33} using the coordinate system for this monograph). C_{11} and C_{33} were determined using the critical angle technique described above for wave propagation in the x and z directions, respectively. C_{55} and C_{13} follow from a similar procedure for quasilongitudinal wave propagation at two angles, say θ_1 and θ_2, with respect to the z axis. Finally, the remaining two orthotropic moduli were calculated from the surface wave velocities in the x and z directions.

CHAPTER 10

Composite Microstructure Characterization

One of the most powerful applications of quantitative ultrasonic techniques to composites lies in the ability to characterize microstructural composition. For polycrystalline metal samples, bulk defects (e.g., cracks, voids, inclusion, etc.) are of primary importance in assessing the structural integrity of completed parts. C-scan techniques have proven to be successful in identifying these defects. While this capability is of importance for composites as well, one must be concerned with a wide array of more subtle defects. Composites achieve their desired properties through engineered microstructures with clearly defined fiber distributions and orientations. Deviations in the microstructure (overcure, undercure, porosity, fiber orientation, fiber segregation, etc.) can result in degraded material properties which can adversely affect performance. As these material anomalies are not associated with large impedance differences from the bulk material, alternative NDE approaches to conventional amplitude discrimination are needed.

Fortunately, there is a great deal more information present in an ultrasonic signal than there is in a conventional C-scan. As we have just seen, anisotropic elastic moduli can be directly measured with a variety of ultrasonic velocity measurement techniques. This capability is coupled with composite micromechanics theory (which relates composite moduli to composition and the properties of the constituent materials) providing a quantitative means of microstructural feature characterization.

COMPOSITE MICROMECHANICS: AN OVERVIEW

The goal of composite micromechanics is the development of the mathematical relations that can be used to predict the moduli of a multiphase material as a function of the properties of its constituents and the composition of the composite. Usually these models are based on idealized representations of the fiber and matrix properties. Typical assumptions include:

- Linearly Elastic [either isotropic (matrix, glass fibers) or transversely isotropic (graphite, Kevlar)]
- Homogeneous
- Perfectly Aligned Fibers
- Randomly Distributed Fibers
- Perfect Bonding between Fiber and Matrix

Micromechanics models are developed through a suitable analysis of a representative volume element (RVE), the basic repeating unit of the composite. The simplest models are based on analyses using "strength of materials" concepts. As an illustration of this approach, consider the RVE shown in Figure 10.1 for a unidirectionally reinforced composite (single ply). If one applies a uniaxial tensile load σ_{11} to the element, we have for the matrix

$$\sigma_m = E_m \epsilon_m$$

for the fiber

$$\sigma_f = E_f \epsilon_f$$

and for the overall composite

$$\sigma_{11} = E_{11} \epsilon_{11}$$

where the subscript 1 denotes the fiber direction. The geometry of the problem requires that

$$\epsilon_f = \epsilon_m = \epsilon_{11}$$

Furthermore, the total force applied to the composite must be distributed between fiber and matrix in the following fashion

$$\sigma_{11} A = \sigma_f A_f + \sigma_m A_m$$

Using the elasticity of the fiber and matrix, we may rewrite this expression as

$$\sigma_{11} A = E_f \epsilon A_f + E_m \epsilon A_m$$

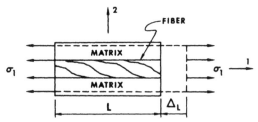

FIGURE 10.1 Geometry for E_{11} calculation—strength of materials approach (after Jones [98]).

Assuming a uniform thickness,

$$V_f = \text{fiber volume fraction} = \frac{\text{fiber volume } (A_f t)}{\text{total volume } (At)} = \frac{\text{fiber area}}{\text{total area}} = \frac{A_f}{A}$$

and

$$V_m = \text{matrix volume fraction} = \frac{\text{matrix volume } (A_m t)}{\text{total volume } (At)} = \frac{A_m}{A}$$

where

$$V_f + V_m = 1.$$

Therefore, we have

$$\sigma_{11} = (E_f V_f + E_m V_m)\epsilon_{11}$$

$$= E_{11}\epsilon_{11}$$

This provides an initial estimate for the stiffness of the composite as a function of composition, as shown in Figure 10.2.

Transverse properties can be obtained in a similar fashion by considering the response of the RVE to a load in the x_2 direction as shown in Figure 10.3. It is readily apparent that, for this case, the applied stress in the fiber and matrix is identical to that of the entire RVE (σ_{22}) so

$$\sigma_2 = \sigma_m = \sigma_f$$

Here, the total width change will be the sum of the width changes in the fiber and matrix or

$$\Delta w = \Delta w_m + \Delta w_f$$

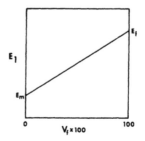

FIGURE 10.2 Strength of materials results for E_{11} (after Jones [98]).

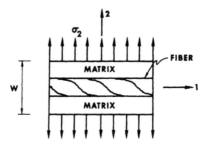

FIGURE 10.3 Transverse properties of composite (after Jones [98]).

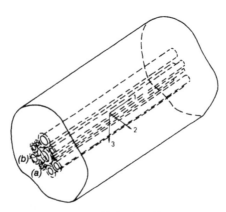

FIGURE 10.4 Composite cylinder assemblage (CCA) model (after Christensen [101]).

156

But,

$$\epsilon_{22} = \frac{\Delta w}{w}, \qquad \epsilon_f = \frac{\Delta w_f}{w_f} = \frac{\Delta w_f}{V_f w}, \qquad \epsilon_m = \frac{\Delta w_m}{w_m} = \frac{\Delta w_m}{V_m w}$$

Substituting these expressions into the preceding equation yields

$$w\epsilon_{22} = V_f w\epsilon_f + V_m w\epsilon_m$$

From the elasticity of the components of the RVE, we have

$$\frac{\sigma_{22}}{E_{22}} = V_f \frac{\sigma_2}{E_f} + V_m \frac{\sigma_2}{E_m}$$

or

$$E_{22} = \frac{E_f E_m}{V_m E_f + V_f E_m}$$

Other moduli of interest can also be modeled in this fashion to yield

$$v_{12} = V_m v_m + V_f v_f$$

$$G_{12} = \frac{G_m G_f}{V_m G_f + V_f G_m}$$

Obviously, several critical assumptions were made in this simplified view of composite behavior. An alternative, elasticity-based approach to the problem has been developed to more realistically represent physical reality. Many of these are based on variational principles and yield upper and lower bounds on moduli rather than exact values. Nevertheless, these corrections to the first-order strength of materials calculations can be important and should be addressed. One of the best ways to model composite behavior is via the composite cylinder assemblage (CCA) model which uses a cylindrical fiber embedded in an annular matrix as the RVE as originally proposed by Hashin and Rosen [99, 100]. In this approach, the fiber radius is taken to be a and matrix radius b (Figure 10.4). The values of a and b are allowed to vary, as long as the ratio a/b remains constant, so that the model will be space filling. Moduli for the RVE are established by subjecting the model to an appropriate strain/stress field and analyzing the resulting deformation using the semi-inverse approach to the solution of the elasticity problem.

To illustrate the process, let us consider E_{fiber}, the Young's modulus for the composite in the fiber direction. The axial symmetry of the problem allows considerable simplification if cylindrical coordinates are used. (Note: For this portion of the analysis, we take the fiber axis to be the z axis of the cylindrical coordinate system.)

$$(u_r)_f = A_f r + \frac{B_f}{r}$$

$$(u_r)_m = A_m r + \frac{B_m}{r}$$

$$(u_\theta)_f = (u_\theta)_m = 0$$

$$(u_z)_f = (u_z)_m = \epsilon z$$

The boundary conditions for this problem will be

(1) $(u_r)_f$ remains finite at the origin (therefore, $B_f = 0$)
(2) Continuity of displacement at fiber matrix interface

$$A_m a + \frac{B_m}{a} = A_f a$$

(3) Continuity of stress at fiber-matrix interface

$$(\sigma_r)_f|_{r=a} = (\sigma_r)_m|_{r=a}$$

(4) Traction-free lateral surface at outer boundary

$$(\sigma_r)_m|_{r=b} = 0$$

Then, one determines the effective Young's modulus in the fiber direction to be

$$E_{\text{fiber}} = \frac{\langle \sigma_{33} \rangle}{\epsilon} = \frac{1}{\pi b^2 \epsilon} \int \int \sigma_{33}(r) \, dA$$

for the individual RVE. The next question that must be answered is whether or not this result is applicable to the collection as a whole. Variational techniques are usually employed to establish a lower bound (using the theorem of minimum potential energy) and an upper bound (using the theorem of minimum complementary energy). Fortunately, for this case, the two bounds coincide and the equation is an exact result with the Young's modulus in the fiber direction given by Christensen [101] as

$$E_{11} = V_f E_f + (1 - V_f) E_m$$

$$+ \frac{4V_f(1 - V_f)(\nu_f - \nu_m)^2 \mu_m}{[(1 - V_f)\mu_m/(K_f + \mu_f/3)] + [V_f \mu_m/(K_m + \mu_m/3)] + 1}$$

for an isotropic fiber in an isotropic matrix. Note that for this equation, the matrix properties are given by Young's modulus E_m, Poisson's ratio ν_m, shear modulus μ_m, and plane strain bulk modulus K_m where only two of the constants are independent with

$$K_m = \frac{E_m}{3(1 - 2\nu_m)}$$

$$\mu_m = \frac{E_m}{2(1 + \nu_m)}$$

with similar relations holding for the fiber properties. Note that the two leading terms in the expression correspond to the strength of materials result obtained earlier, with the final term being a correction. A similar result can be obtained for the Poisson ratio ν_{12} using this approach as

$$\nu_{12} = V_f \nu_f + (1 - V_f) \nu_m$$

$$+ \frac{V_f(1 - V_f)(\nu_f - \nu_m)[\mu_m/(K_m + \mu_m/3) - \mu_m/(K_f + \mu_f)/3]}{[(1 - V_f)\mu_m/(K_f + \mu_f/3) + [V_f \mu_m/(K_m + \mu_m/3)] + 1}$$

The remaining moduli are given by

$$K_{23} = K_m$$

$$+ \frac{\mu_m}{3} \frac{V_f}{1 \Big/ \Big[K_f - K_m + \frac{1}{3}(\mu_f - \mu_m) \Big] + (1 - V_f) \Big/ \Big(K_m + \frac{4}{3}\mu_m \Big)}$$

$$\frac{\mu_{12}}{\mu_m} = \frac{\mu_f(1 + V_f) + \mu_m(1 - V_f)}{\mu_f(1 - V_f) + \mu_m(1 + V_f)}$$

$$\frac{\mu_{23}}{\mu_m} = 1 + \frac{V_f}{[\mu_m/(\mu_f - M_m)] + \Big(K_m + \frac{7}{3}\mu_m \Big) \Big/ \Big(2K_m + \frac{8}{3}\mu_m \Big)}$$

where for μ_{23}, the transverse shear modulus, the upper and lower bounds do not coincide and only the lower bound is presented for simplicity. To recover the stiffness elements for the composite, the following equations can be used

$$C_{11} = E_{11} + 4\nu_{12}^2 K_{23}$$

$$C_{12} = 2K_{23}\mu_{12}$$

$$C_{22} = \mu_{23} + K_{23}$$

$$C_{23} = -\mu_{23} + K_{23}$$

$$C_{66} = \mu_{12}$$

Hashin has further refined these equations to allow for anisotropic fiber properties, as for graphite Kevlar and boron. These results are similar to those just presented, and are found in Reference 102.

Table 10.1. Material Properties of Reinforcing Fibers from Micromechanics Model and Ultrasonic Measurements (after Smith [103]).

Plate		Static Young's Modulus E_L (10^6 psi)	Ultrasonic Modulus (10^6 psi)		Ultrasonic Poisson Ratios		
			E_L	E_T	ν_{LT}	ν_{TL}	ν_{TT}
	Graphite fibers						
1	WYB	8.2	7.6	2.3	0.15	0.05	0.28
2	"Thornel" 25	22.4	19.8	1.53	0.23	0.02	0.32
3	"Thornel" 40	42.6	36.2	1.21	0.23	0.01	0.39
4	"Thornel" 50	51.1	47.2	1.10	0.41	0.01	0.45
5							
6	"Thornel" 50S	59.7	55.1	1.05	0.31	0.005	0.48
7							
8	"Thornel" 50S	61.6	55.8	1.03	0.37	0.007	0.47
9	"Thornel" 75S	81.2	78.3	0.82	0.39	0.004	0.56
	Carbon fibers						
10	VYB	5.9	4.9	5.14	0.37	0.40	−0.03
11	Experimental PAN	30.0	27.9	2.45	0.30	0.03	0.38
12	HTS	42.6	35.5	1.85	0.28	0.02	0.33
13	"Thornel" 400	30.0	25.5	2.41	0.23	0.02	0.41

EXPERIMENTAL RESULTS

Smith [103] was among the first to realize the connection between composite micromechanics and ultrasonic test capabilities. His experimental setup has been described previously in this monograph. Of particular interest in Smith's work is the use of micromechanics principles to study the effects of constituent properties (particularly fiber modulus) on composite properties and therefore ultrasonic velocities. Using ultrasonic data for the composite and known resin properties, Smith was able to extract the material properties for a variety of carbon and graphite reinforcing fibers. Typical results are shown in Table 10.1. Recently, Datta and co-workers [104] used a similar approach for taking elastic property measurements of graphite fibers. Both type I and type II fibers were analyzed.

These results are particularly useful as, while the longitudinal properties of these fibers are relatively easy to measure experimentally, the transverse properties are quite difficult to characterize due to their small diameter. Gieske and Allred [105] used a composite micromechanics to study the effects of composition (here fiber volume fraction) on a unidirectional basic-aluminum composite. Micromechanics predictions of moduli compared quite favorably with ultrasonically measured values for all of the moduli considered. Results are presented in Figure 10.5. Dean and Turner [106] also used a similar micromechanics-based approach to study the effects of composition on the elastic moduli of graphite fiber composites. Typical results are shown in Figure 10.6.

Kriz and Stinchcomb [107] later verified the accuracy of the composite micromechanics of Hashin [102] via extensive comparison with ultrasonic test measurements. As shown in Figures 10.7–10.8, the agreement is in most cases good, with the principal exception being the Poisson ratio transverse to the fibers. As pointed out earlier, this property is the most difficult modulus to measure as it is highly sensitive to experimental uncertainties.

Results such as these were the basis for the work of Reynolds and Wilkinson [108] who realized the importance of micromechanics in the quantitative characterization of composite material microstructure.

In their work, it was recognized that the presence of long, cylindrical reinforcing fibers will have a directionally dependent effect on elastic properties, but that the contribution from spherical pores (providing they are randomly distributed) will be directionally independent. Hence, they will have a different effect on longitudinal and shear (polarized) waves. This means each combination of porosity and fiber volume fraction is associated with a unique longitudinal velocity/shear velocity pair, as illustrated in Figure 10.9.

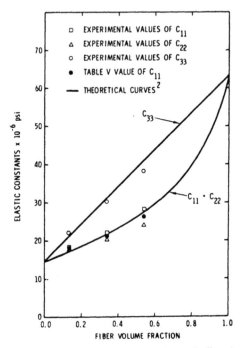

Courtesy Society for Experimental Mechanics, Inc.

FIGURE 10.5 Comparison of micromechanics predictions with ultrasonic test results for B-Al composite (after Gieske and Allred [105]).

Reynolds and Wilkinson then measured the longitudinal velocity and shear velocity at a given point and used graphical means of establishing the fiber volume fraction and porosity present. It should be pointed out that this approach requires precise knowledge of the properties (density, elastic moduli) of the microconstituents. Also, the Reynolds and Wilkinson approach suffers from several critical drawbacks in that it uses local, contact measurements (hence is unsuitable for scanning), employs a graphical (hence slow and inaccurate) solution technique, and can only be applied to unidirectionally reinforced composites (not practical laminates). In addition, no experimental data are presented to corroborate their findings. Nonetheless, their work is conceptually sound and represents an important pioneering effort in a critical discipline. Recent work in the area has attempted to address some of these shortcomings.

Kline and co-workers [109–111] have developed automated procedures for calculating fiber volume fraction and porosity from the measured ultrasonic data. Initially, this was achieved through a mesh refinement procedure illustrated in Figure 10.10. With this approach, predicted longitudinal and shear (contact) velocity pairs are calculated for relatively coarse increments in fiber volume fraction and porosity. The closest pair to the measured data is then identified. A fine mesh is then constructed in this vicinity and the process is repeated until a sufficiently accurate measure of fibers volume fraction and porosity is obtained. Using this algorithm, Gruber et al. [112] extended this approach to a scanning configuration to map out property variations. While restricted to unidirectionally reinforced samples and the plane of transverse isotropy (parallel to the fibers), the study is nonetheless important as it demonstrates the ability of this approach to image local variations in material properties in composites. Also of interest in this

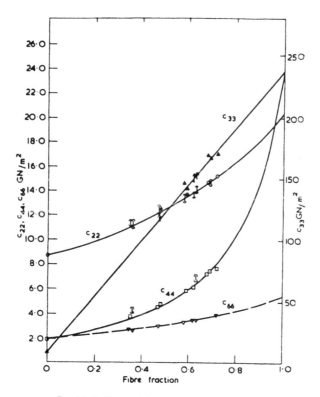

FIGURE 10.6 Comparison of micromechanics predictions with ultrasonic test results for *Gr/Ep* (after Dean and Turner [106]).

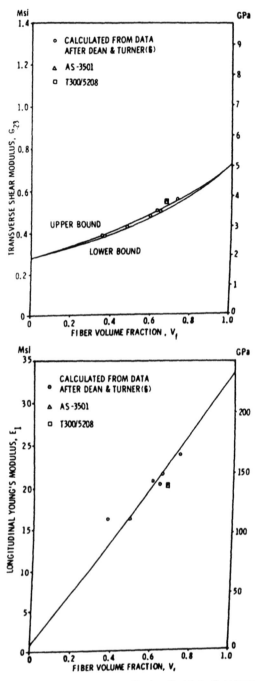

Courtesy Society for Experimental Mechanics, Inc.

FIGURE 10.7 Comparison of micromechanics predictions with ultrasonic test results for *Gr/Ep* (after Kriz and Stinchcomb [107]).

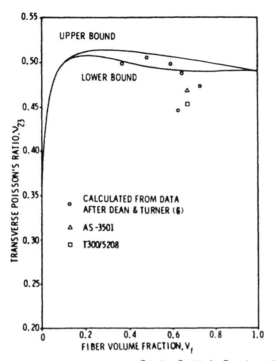

FIGURE 10.8 Comparison of micromechanics predictions (for transverse Poisson's ratio) with ultrasonic test results for *Gr/Ep* (after Kriz and Stinchcomb [107]).

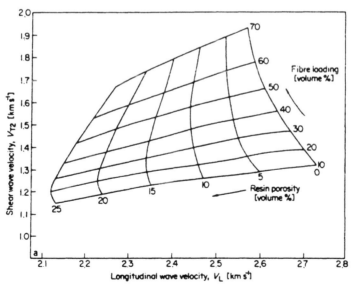

FIGURE 10.9 Effect of fiber volume fraction and porosity on longitudinal and shear velocities (after Reynolds and Wilkinson [108]).

165

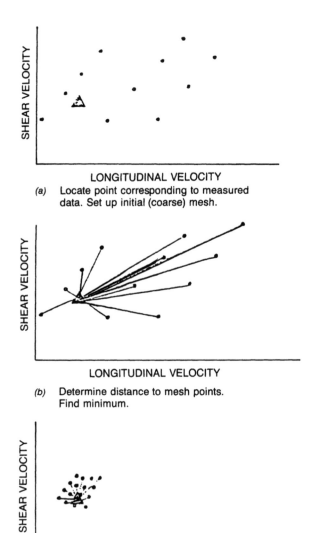

(a) Locate point corresponding to measured data. Set up initial (coarse) mesh.

(b) Determine distance to mesh points. Find minimum.

FIGURE 10.10 Mesh refinement procedure (after Kline [109]).

166

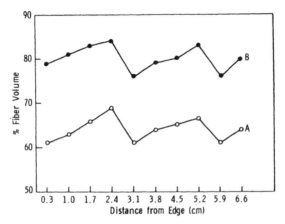

FIGURE 10.11 Comparison of ultrasonic test results with quantitative image analysis–qualitative agreement (after Gruber et al. [112]).

study is an important caveat regarding quantitative microstructural characterization of composites using micromechanics. This method requires accurate knowledge of the densities and elastic properties of the constituents. Without this critical information, errors can occur. This is illustrated in Figure 10.11 where material microstructure parameters as determined via ultrasonics and image analysis are compared. Note the qualitative, but not quantitative agreement between the two studies. The principal reason for the lack of quantitative agreement between the two results is inaccurate matrix property data. Composite matrix properties are very sensitive to the degree of cure during processing. Hence, without some knowledge of the cure state, errors may be introduced. However, when the actual matrix properties are well characterized, good agreement is obtained both qualitatively and quantitatively, as in Figure 10.12.

Recently, Kline and Adams have extended these results to the scan of laminated composite structures [110]. Most efforts in this area are based on common micromechanics assumptions, i.e., linear elasticity, perfect bonding between fiber and matrix, perfect fiber alignment, isotropic matrix, isotropic or transversely isotropic fibers, known properties of constituent materials, etc. In order to extend these results to laminated materials, two additional simplifying assumptions were employed.

(1) The plies are perfectly aligned with a known stacking sequence.
(2) The composition of each ply in the laminate is identical.

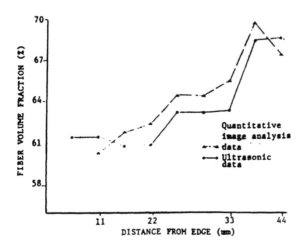

FIGURE 10.12 Comparison of ultrasonic test results with quantitative image analysis–qualitative and quantitative agreement (after Kline [109]).

With these assumptions, the problem is tractable using techniques similar to those previously developed for unidirectionally reinforced media. Since scanning was desired, mode conversion was employed although pure modes will no longer be generated and energy flux deviation from the wave normal must be considered. These effects were taken into account in the software.

The problem was posed as a coupled set of nonlinear equations. The transit time for a given wave mode will, in general, be a function of the elastic moduli of the medium, the density, the direction of propagation, the layup sequence of the composite, and the path traversed in the medium, say

$$TT_i = TT_i(C_{ij}, \rho, \theta)$$

Also, from composite micromechanics, the elastic moduli of the composite can be determined from the properties of the fiber and resin systems, the number of reinforcing fibers present and the porosity of the media

$$(C_{ij})_{\text{composite}} = C_{ij}[(C_{ij})_{\text{fiber}}, (C_{ij})_{\text{resin}}, V_{\text{fiber}}, V_{\text{porosity}}]$$

As they sought to determine the parameters that characterize specimen composition (here fiber volume fraction and porosity) from acoustic experiments, two independent equations were needed. Hence, two measurements of the transit time for two independent waves were required. In thin composite laminates, one choice is obvious—the through thickness longitudinal wave. The choice for the second wave, however, is less clear. Shear wave propagation with contact transducers is possible, but the need for a high-viscosity coupling agent precludes scanning in most cases. Mode conversion was therefore used, with suitable care exercised due to energy flux deviation from the wave normal and refraction phenomena at each ply boundary. Here, they chose to use a quasilongitudinal wave generated by an obliquely incident longitudinal wave in water. This choice has the advantage of being readily isolated from other waves, as it will be the first arrival in a through transmission experiment. The choice of inspection angle is based on two competing factors: localized inspection area (small angle) and the ability to resolve differences in wave speed due to the presence of reinforcing fibers (large angle). Figure 10.13 illustrates this problem for two oblique inspection angles in a cross-ply composite. Clearly, at 5°, resolution capability for fiber volume fraction is limited in comparison to that for the 7° inspection. However, this means that a larger volume of material is studied with each measurement.

A comparison of ultrasonic results for twelve samples cut from the plate with acid digestion is shown in Figure 10.14. Clearly, the agreement between the two approaches is quite good with most readings within ±2%, the inherent accuracy of the acid digestion technique.

This research group has also demonstrated that this same approach can be used to automatically scan composite prepreg materials on-line. The test method utilizes a set of transducers imbedded inside a rolling mill. The mill is filled with water to act as coupling in a water-wheel type device. A central transducer is used to generate a longitudinal

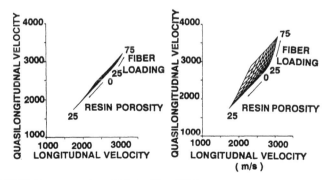

FIGURE 10.13 Measurement sensitivity for 5° and 7° inspection angles (after Kline and Adams [110]).

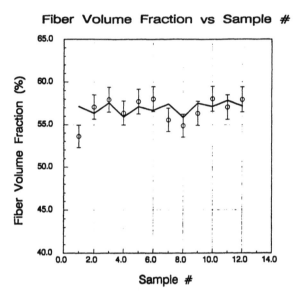

FIGURE 10.14 Comparison of ultrasonic test results with acid digestion measurements (after Kline and Adams [110]).

wave at normal incidence while a set of two additional transducers are used in a pitch-catch arrangement (perpendicular to the reinforcing fibers) to generate the second (pure mode transverse) mode required for microstructure characterization. To simulate a production environment, a computer-controlled composite prepreg rolling mill (Figure 10.15) was developed. Prepregs were rolled, while being scanned by ultrasonic transducers set up in pitch catch mode inside the rolling mill. A software interface has been developed between the rolling mill and a personal computer to control the speed, acceleration of the rollers and the distance traveled by the prepreg material. To ensure accurate transit time measurement in the thin prepreg layer, the analytic signal magnitude technique was used. The algorithm developed by Kline and Adams [110] was modified to account for the experimental geometry of the experiment. Fiber volume fraction and the porosity of the prepreg were determined from the material properties and the measured wave velocities as the material was processed. Results are shown in Figure 10.16.

Belasubramanian and co-workers [113] at Drexel have also used a coupled micromechanics/ultrasonics approach to study microstructural features such as fiber misalignment and porosity in unidirectionally reinforced composite media. This group utilizes the leaky Lamb wave approach to measure the dispersion curves for the materials. Typical results for these two material anomalies are shown in Figure 10.17. Through the use of an iterative scheme and composite micromechanics, it is possible to adjust the predicted dispersion curves to bring them into agreement with the measured curves and, hence, to determine microstructure.

Porosity effects have also been extensively studied by Hsu and co-workers [114–116] using both velocity and attenuation-based measurements. Measurements of ultrasonic velocity were made over a wide frequency range to characterize dispersion in porosity composite laminates. Both woven as well as quasi-isotropic composites were studied. Results from these samples are presented in Figure 10.18 and qualitatively show the effect of void content on dispersion. Frequency-dependent attenuation was also found to correlate well with porosity content. Attenuation in composites is often found to vary linearly with frequency as shown in Figure 10.19. The slope of the attenuation vs. frequency curve has been found to correlate well with porosity (Figure

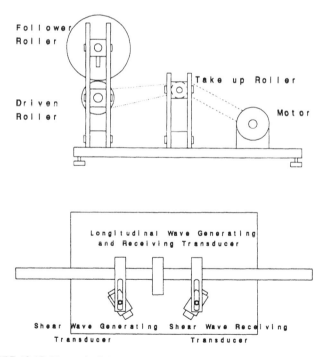

FIGURE 10.15 Ultrasonically instrumented rolling mill (after Kline and Kulathu [111]).

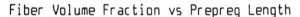

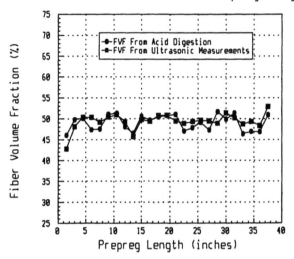

FIGURE 10.16 Prepreg inspection results: Comparison of ultrasonic test results with acid digestion measurements (after Kline and Kulathu [111]).

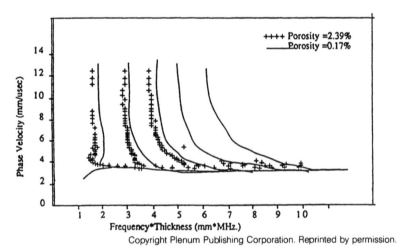

Copyright Plenum Publishing Corporation. Reprinted by permission.

FIGURE 10.17 LLW test results for composite porosity (after Belasubramanian et al. [113]).

172

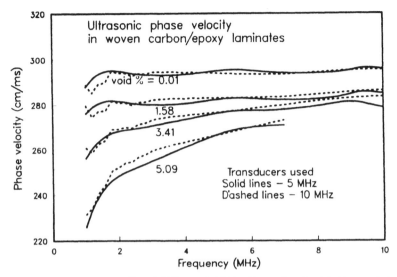

FIGURE 10.18 Effect of void content on acoustic wave dispersion (after Hsu et al. [115]).

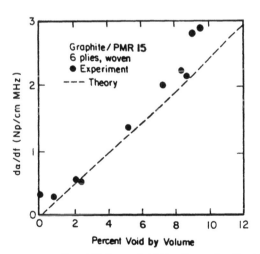

FIGURE 10.19 Correlation of slope of attenuation vs. frequency curves with porosity (after Hsu et al. [116]).

173

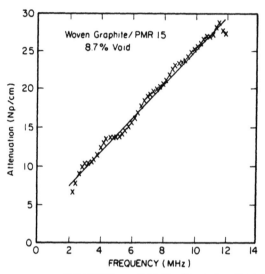

FIGURE 10.20 Variation of attenuation with frequency in composites (after Hsu et al. [116]).

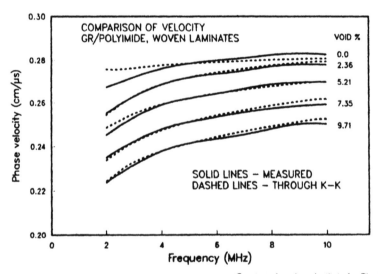

FIGURE 10.21 Comparison of measured velocity in graphite–polyimide composites with porosity and calculated velocity using Kramers–Kroenig relation and $\omega_0 = 5$ MHz.

174

10.20). Hsu et al. have used the following empirical relationship to describe this phenomenon

$$\text{Void Content } (\%) = K \frac{d\alpha}{df}$$

where the constant K depends on the morphology of the pores. This group has also found the local form of the Kramers–Kroenig relationships [117] to hold for composite materials (Figure 10.21) as one would expect in a causal system.

REFERENCES

1. Green, R. E., Jr. 1973. *Ultrasonic Investigation of Mechanical Properties, Treatise on Materials Science, Vol. 3.* H. Hermann, ed. New York: Academic Press.
2. Kriz, R. and H. Ledbetter. 1985. "Elastic Representation Surfaces of Unidirectional Graphite-Epoxy Composites," in *Recent Advances in Composites*, ASTM STP864. J. Vinson and M. Taya, eds. Philadelphia: ASTM, pp. 661–675.
3. Henneke, E. G. 1972. "Reflection-Refraction of a Stress Wave at a Plane Boundary Between Anisotropic Media," *J. Acoust. Soc. Am.*, 51:210–217.
4. Fedorov, F. I. 1968. *Theory of Elastic Waves in Crystals.* New York: Plenum Press.
5. Green, R. E. Jr. 1973. *Ultrasonic Investigation of Mechanical Properties, Treatise on Materials Science, Vol. 3.* H. Hermann, ed. New York: Academic Press.
6. Henneke, E. G. and G. L. Jones. 1976. "Critical Angle for Reflection at a Solid-Liquid Interface in Single Crystals," *J. Acoust. Soc. Am.*, 59:205–206.
7. Henneke, E. G. 1972. "Reflection-Refraction of a Stress Wave at a Plane Boundary Between Anisotropic Media," *J. Acoust. Soc. Am.*, 51:210–217.
8. Musgrave, M. J. P. 1970. *Crystal Acoustics.* San Francisco: Holden-Day.
9. Rokhlin, S. I. and D. Chimenti. 1989. "Reconstruction of Elastic Constants from Ultrasonic Reflectivity Data in a Fluid Coupled Composite Plate," *Proceedings QNDE*, 9:1411–1419.
10. Graff, K. 1975. *Wave Motion in Elastic Solids.* Columbus, OH: Ohio State.
11. Musgrave, M. J. P. 1970. *Crystal Acoustics.* San Francisco: Holden-Day.
12. Hosten, B., M. Deschamps, and B. Tittmann. 1987. "Inhomogeneous Wave Generation and Propagation in Lossy Anisotropic Solids Application to the Characterization of Viscoelastic Composite Materials," *J. Acoust. Soc. Am.*, 82:1763–1770.
13. Rayleigh, J. W. S. 1887. "On Waves Propagated Along the Plane Surface of an Elastic Solid," *Proc. Lond. Math. Soc.*, 17:4–11.
14. Lamb, H. 1917. "On Waves in an Elastic Plate," *Proc. R. Soc. A*, A93:114–128.
15. Love, A. E. H. 1911. *Some Problems of Geodynamics.* Cambridge University Press.
16. Stoneley, R. 1924. "Elastic Waves at the Surface of Separation of Two Solids," *Proc. R. Soc. A*, A106:416–428.

177

17. Ekstein, H. E. 1945. "High Frequency Vibration of Thin Crystal Plates," *Phys. Rev.*, 68:11–23.

18. Pao, Y.-H. and R. K. Kaul. 1974. "Waves and Vibrations in Isotropic and Anisotropic Plates," in *R. D. Mindlin and Applied Mechanics.* New York: Pergamon, pp 149–195.

19. Farnell, G. W. 1970. "Properties of Elastic Surface Waves," in *Physical Acoustics,* pp 109–106.

20. Buchwald, V. T. 1961. "Rayleigh Waves in Transversely Isotropic Media," *Quant. J. Mech. Appl. Math.*, 14:293–317.

21. Lim, T. C. Ph.D. Thesis, McGill University (unpublished).

22. Stoneley, R. 1955. *Proc. R. Soc. Lond. A*, A2323:447–458.

23. Synge, R. 1957. *J. Math. Phys.*, 35:323.

24. Royer, D. and E. Dieulesaint. 1984. "Rayleigh Wave Velocity and Displacement in Orthorhombic, Tetragonal, Hexagonal and Cubic Crystals," *J. Acoust. Soc. Am.*, 76:1438–1444.

25. Rose, J., A. Pilarski, and Y. Huang. 1990. "Surface Wave Utility in Material Characterization," *Res. Nondestr. Eval.*, 1:247–265.

26. Green, W. A. 1982. "Bending Waves in Strongly Anisotropic Elastic Plates," *J. Mech. Appl. Math.*, XXXV, pp 485–507.

27. Green, W. A. and D. Milosavljevic. 1985. "Extensional Waves in Strongly Anisotropic Elastic Plates," *Intl. J. Solids Structures*, 21:343–353.

28. Kline, R. A., M. M. Doroudian, and C. P. Hsiao. 1989. "Plate Wave Propagation in Transversely Isotropic Materials," *J. Comp. Materials*, 23:505–533.

29. Chimenti, D. and A. Nayfeh. 1990. "Ultrasonic Reflection and Guided Wave Propagation in Biaxially Laminated Composite Plates," *J. Acoust. Soc. Am.*, 87:1409–1415.

30. Wang, W. and S. I. Rokhlin. 1990. "Measurements of Elastic Constants of Metal Matrix Ceramic Composite using Ultrasonic Plate Mode Antiresonance," in *Nondestructive Characterization of Materials.* R. Green and C. Rudd, eds. (in press).

31. Mal, A. K., M. Karim, and Y. Bar-Cohen. 1990. "Determination of the Properties of Composite Interfaces by an Ultrasonic Method," *Material Science and Engineering*, A126:155–163.

32. Bar-Cohen, Y. and A. K. Mal. 1988. "Leaky Lamb Waves in Composites using Pulses," *Proceedings QNDE*, 8.

33. Bar-Cohen, Y. and A. K. Mal. 1989. "Leaky Lamb Waves in Multiorientation Composite Laminates," *Proceedings QNDE*, 9:1419–1424.

34. Bar-Cohen, Y. and A. K. Mal. "Characteristics of Composite Laminates using Combined LLW and PBS Methods," *Proceedings QNDE*, in press.

35. Dayal, V. and V. K. Kinra. 1989. "Leaky Lamb Waves in an Anisotropic Plate: An Exact Solution and Experiments," *J. Acoust. Soc. Am.*, 85:2268–2276.

36. Rokhlin, S. I. and D. Chimenti. 1989. "Reconstruction of Elastic Constants from Ultrasonic Reflectivity Data in a Fluid Coupled Composite Plate," *Review of Progress in QNDE*, 9:1411–1419.

37. Farnell, G. W. and E. L. Adler. 1972. "Elastic Wave Propagation in Thin Layers," *Physical Acoustics*, 9:35–127.

38. Bouden, N. and S. Datta. 1989. "Rayleigh and Love Waves in Cladded Anisotropic Medium," *Review of Progress in QNDE,* 10:1337–1344.

39. Truell, R., C. Elbaum, and B. Chick. 1969. *Ultrasonic Methods in Solid State Physics.* New York: Academic Press.

40. Mason, W. P. and H. Bommel. 1956. "Ultrasonic Attenuation at Low Temperature for Metals in Normal and Superconducting States," *J. Acoust. Soc. Am.,* 28:930–934.

41. Egle, D. M. 1980. "Using the Acoustoelastic Effect to Measure Stress in Plates," UCRL-52914, Lawrence Livermore Laboratory.

42. Castagnede, B., J. Roux, and B. Hosten. 1989. "Correlation Method for Normal Mode Tracking in Anisotropic Media using an Ultrasonic Immersion System," *Ultrasonics,* 22:280–287.

43. Chang, F., J. Couchman, and B. Yee. 1974. "Ultrasonic Resonance Measurements of Sound Velocity in Thin Composite Laminates," *J. Composite Materials,* 8:356–363.

44. Kinra, V. and V. Dayal. 1988. "A New Technique for Ultrasonic Nondestructive Evaluation of Thin Specimens," *Experimental Mechanics,* 28:288–297.

45. Iyer, V., S. Hanneman, and V. Kinra. 1990. "Use of Phase Spectra for Ultrasonic NDE of Sub-Half Wavelength Composite Laminates," *Advances in Composites, ASTM,* pp 63–105.

46. Heyser, R. G. 1969. "Determination of Loud Speaker Arrival Times," *Audio Eng. Soc.,* 14:902–905.

47. Gammel, P. M. 1981. "Analogue Implementation of Analytic Signal Magnitude for Pulse-Echo Systems," *Ultrasonics,* 19:279–283.

48. Bar-Cohen, Y. 1985. "Ultrasonic NDE of Composites—A Review," in *Proceedings, Solid Mechanics Research for QNDE,* Evanston, IL.

49. Chimenti, D. and R. Martin. "Nondestructive Evaluation of Composite Laminates," *Ultrasonics,* in press.

50. Rokhlin, S. I., C. Y. Wu, and L. Wang. 1990. "Application of Coupled Ultrasonic Plate Modes for Elastic Constant Reconstruction in Anisotropic Composites," in *Review of Progress in QNDE,* 9:1403–1411.

51. Elden, L. and L. Wittmeyer-Koch. 1988. *Numerical Analysis: An Introduction.* Boston: Academic Press.

52. Marquardt, D. W. 1963. "An Algorithm for Least Square Estimation of Nonlinear Parameter," *J. Soc. Ind. Appl. Math.,* 11:431–448.

53. Kline, R. A., S. K. Sahay, and E. I. Madaras. 1990. "Ultrasonic Modulus Determination in Composite Media," *Review of Progress in QNDE,* in press.

54. Wooh, S. and I. Daniel. 1990. "Mechanical Characterization of a Unidirectional Composite by Ultrasonic Methods," *Review of Progress in QNDE,* in press.

55. Mignogna, R., N. Batra, and H. Chaskelis. 1990. "Wave Propagation in Nonsymmetry Directions in Anisotropic Media," *Review of Progress in QNDE,* in press.

56. Karim, M., A. Mal, and Y. Bar-Cohen. 1990. "Inversion of Leaky Lamb Wave Data by Simplex Algorithm," *J. Acoust. Soc. Am.,* 88:482–491.

57. Caceci, M. and W. Cacheris. 1984. "Fitting Curves to Data," *Byte,* pp 340–362.

58. Pearson, L. H. and W. J. Murri. 1986. "Measurement of Ultrasonic Wavespeeds

in Off-Axis Directions of Composite Materials," *Review of Progress in QNDE*, 5: 1093–1101.

59. Mignogna, R. B. 1989. "Ultrasonic Determination of Elastic Constants from Oblique Angles of Incidence in Non-Symmetry Planes," *Review of Progress in QNDE*, in press.

60. Kline, R., S. Sahay, Y. Wang, R. Adams, R. Kulathu, and R. Mignogna. "Phase and Group Velocity Considerations for Dynamic Modulus Measurement in Anisotropic Media," *Ultrasonics*, in press.

61. Tauchert, T. and Guselzu. 1972. "An Experimental Study of Dispersion of Shear Waves in a Fiber Reinforced Composite," Paper 71-APM-27, *J. Appl. Mech.*, 39: 98–102.

62. Grabel, J. and J. Cost. 1972. "Ultrasonic Velocity Measurement of Elastic Constants of 1-Al₃Wᵢ Unidirectionally Solidified Eutectic," *Metall. Trans.*, 3:1973–1977.

63. Sachse, W. 1974. "Measurement of the Elastic Moduli of Continuous Filament and Eutectic Composite Materials," *J. Composite Materials*, 8:378–390.

64. Dean, G. and P. Turner. 1973. "The Elastic Properties of Carbon Fibers and Their Composites," *Composites*, 3:174–180.

65. Zimmer, J. E., and J. R. Cost. 1970. "Determination of Elastic Constants of a Uni-directional Fiber Composite Using Ultrasonic Velocity Measurements," *J. Acoust. Soc. Am.*, 47:795–803.

66. Kriz, R. and W. Stinchcomb. 1979. "Elastic Moduli of Transversely Isotropic Graphite Fibers and Their Composites," *Experimental Mechanics*, 19:41–49.

67. Jeong, H., D. Hsu, R. Shannon, and P. Liaw. 1990. "Elastic Moduli of Silicon Carbide Particulate Reinforced Aluminum Metal Matrix Composites," *Review of Progress in QNDE*, 9.

68. Markham, M. F. 1970. "Measurement of the Elastic Constants of Fibre Composites by Ultrasonics," *Composites*, 1:145–149.

69. Pearson, L. H. and W. J. Murri. 1986. "Measurement of Ultrasonic Wavespeeds in Off-Axis Directions of Composite Materials," *Review of Progress in QNDE*, 5: 1093–1101.

70. Smith, R. E. 1972. "Ultrasonic Elastic Constants of Carbon Fibers and Their Composites," *J. Appl. Phys.*, 43:2555–2561.

71. Gieske, J. and R. Allred. 1974. "Elastic Constants of B-Al Composites by Ultrasonic Velocity Measurements," *Exp. Mechanics*, 14:158–165.

72. Hosten, B. and B. Castagnede. 1983. "Optimization du calculdes Constantes Elastiques á Partiv des Mesures de Vitesses d'une onde Ultrasonore," *C.R. Acad. Sc. Paris, Série II*, 8:297–300.

73. Hosten, B., A. Barrot, and J. Roux. 1983. "Interférométrie Numerique Ultrasonore Pour la Determination de la Matrice de Raideur des Matériaux Composites," *Acoustica*, 53:212–217.

74. Hosten, B. and B. Tittmann. 1986. "Elastic Anisotropy of Carbon-Carbon Composites During the Fabrication Process," *IEEE Transactions on Ultrasonics, Ferroelectrics and Frequency Control*, UFFC-34, pp. 1061–1063.

75. Hosten, B., M. Deschamps, and B. Tittmann. 1987. "Inhomogeneous Wave Generation and Propagation in Lossy Anisotropic Solids Application to the Characterization of Viscoelastic Composite Materials," *J. Acoust. Soc. Am.*, 82:1763–1770.

76. Kline, R. A. and Z. T. Chen. 1988. "Ultrasonic Technique for Global Anisotropic Property Measurement in Composite Material," *Mat. Eval.*, 46:986–992.

77. Kline, R. A. 1990. "Quantitative NDE of Advanced Composites using Ultrasonic Velocity Measurements," *J. Eng. Mat. and Tech.*, 112:218–222.

78. Kline, R. A., G. Cruse, A. G. Striz, and E. I. Madaras. 1989. "Analysis of Carbon-Carbon Composites using an Integrated NDE/Finite Element Approach," *1989 Ultrasonic Symposium Proceedings IEEE*, pp 1171–1175.

79. Kline, R. A., S. K. Sahay, and E. I. Madaras. 1990. "Ultrasonic Modulus Determination in Composite Media," *Review of Progress in QNDE*, in press.

80. Hsu, D. K. and F. J. Margetan. June 1990. "Quantitative Analysis of Oblique Echoes in Thick Composites using the Slowness Surfaces," *Proceedings American Society for Composites East Lansing, MI*, pp 945–954.

81. Rokhlin, S. and D. Chimenti. 1989. "Reconstruction of Elastic Constants from Ultrasonic Reflectivity Data in a Fluid-Coupled Composite Plate," *Review of Progress in QNDE*, 9:1411–1419.

82. Dayal, V. and V. Kinra. 1989. "Leaky Lamb Waves in an Anisotropic Plate: An Exact Solution and Experiments," *J. Acoust. Soc. Am.*, 85:2268.

83. Xu, P., A. Mal, and Y. Bar-Cohen. 1990. "Inversion of Leaky Lamb Wave Data to Determine Cohesive Properties of Bonds," *Int. J. Eng. Sci.*, 28:331–346.

84. Karim, M., A. Mal, and Y. Bar-Cohen. 1990. "Inversion of Leaky Lamb Wave Data by Simplex Algorithm," *J. Acoust. Soc. Am.*, 88:482–491.

85. Rokhlin, S. I., C. Y. Wu, and L. Wang. 1990. "Application of Coupled Ultrasonic Plate Modes for Elastic Constant Reconstruction of Anisotropic Composites," *Review of Progress in QNDE*, 9:1403–1411.

86. Wang, W. and S. Rokhlin. June 1990. "Measurements of Elastic Constants of Metal Matrix Ceramic Composites using Ultrasonic Plate Mode Antiresonances," *4th Int. Symposium on Nondestructive Characterization of Materials, Annapolis* (in press).

87. Payton, R. G. 1983. *Elastic Wave Propagation in Transversely Isotropic Media.* Hingham, MA: M. Nijhoff.

88. Rose, J. L., K. Balasubramaniam, and A. Tuerdokhlebou. 1989. "A Numerical Integration Green's Function Model for Ultrasonic Field Profiles in Mildly Anisotropic Media," *J. Nondest. Eval.*, 8:165–179.

89. Doyle, P. 1989. "Calculation of Ultrasonic Surface Waves from an Extended Thermoelastic Source," *J. Nondest. Eval.*, 8:147–164.

90. Doyle, P. and C. Scala. 1990. "Ultrasonic Measurement of Elastic Constants for Composite Overlays," *Review of Progress in QNDE*, in press.

91. Sachse, W., B. Castagnede, I. Grabel, K. Kim, and R. Weaver. 1990. "Recent Developments in Quantitative Ultrasonic NDE of Composites," *Ultrasonics*, 28: 97–104.

92. Castagnede, B. and W. Sachse. 1988. "Optimized Determination of Elastic Constants in Anisotropic Solids from Wavespeed Measurements," *Review of Progress in QNDE*, 8:1855–1862.

93. Sachse, W., A. Every, and M. Thompson. June 1990. "Impact of Laser Pulses on Composites Materials," *Proceedings American Society for Composites, East Lansing, MI*, pp 51–62.

94. Rokhlin, S. and L. Wang. 1989. "Critical Angle Measurement of Elastic Constants in Composite Material," *J. Acoust. Soc. Am.,* 86:1876–1881.

95. Henneke, E. G. 1972. "Reflection and Refraction of a Stress Wave on a Plane Boundary Between Anisotropic Media," *J. Acoust. Soc. Am.,* 51:210–217.

96. Henneke, E. G. and G. Jones. 1976. "Critical Angle for Reflection at Liquid Solid Interface in Single Crystals," *J. Acoust. Soc. Am.,* 59:204–205.

97. Pilarski, A. and J. Rose. 1989. "Use of Subsurface Longitudinal Waves in Composite Materials Characterization," *Ultrasonics,* 27:226–233.

98. Jones, R. 1978. *Mechanics of Composite Materials.* New York: Hemisphere.

99. Hashin, Z. and B. W. Rosen. 1964. "The Elastic Moduli of Fiber Reinforced Materials," *J. Appl. Mech.,* 31:223–232.

100. Hashin, Z. 1965. "On Elastic Behavior of Fibre Reinforced Materials of Arbitrary Transverse Phase Geometry," *J. Mech. Phys. Solids,* 13:119–134.

101. Christensen, R. 1979. *Mechanics of Composite Materials.* New York: Wiley Interscience.

102. Hashin, Z. 1979. "Analysis of Properties of Fiber Composites with Anisotropic Constituents," *J. Appl. Mech.,* 46:543–549.

103. Smith, R. E. 1972. "Ultrasonic Constants of Carbon Fibers and Their Composites," *J. Appl. Phys.,* 43:2555–2561.

104. Datta, S., H. Ledbetter, and T. Kyono. 1970. "Graphite Fiber Elastic Constants: Determination from Ultrasonic Measurements on Composite Materials," *Review of Progress in QNDE,* pp 1481–1488.

105. Gieske, J. and R. Allred. 1974. "Elastic Constants of B-Al Composites by Ultrasonic Velocity Measurements," *Experimental Mechanics,* 14:158–166.

106. Dean, G. and P. Turner. 1973. "The Elastic Properties of Carbon Fibers and Their Composites," *Composites,* 3:174–180.

107. Kriz, R. and W. Stinchcomb. 1979. "Elastic Moduli of Transversely Isotropic Graphite Fibers and Their Composites," *Experimental Mechanics,* 19:41–49.

108. Reynolds, W. and S. Wilkinson. 1978. "The Analysis of Fiber Reinforced Porous Composite Materials by the Measurement of Ultrasonic Wave Velocities," *Ultrasonics,* 16:159–163.

109. Kline, R. 1988. "Ultrasonic Measurement of Fiber Volume Fraction and Porosity in Composites," in *Proceedings ENTEC, Dallas, TX.*

110. Kline, R. and R. Adams. 1991. "NDE of Composite Laminates," to appear in *Nondestructive Characterization of Materials,* C. Rudd and R. E. Green, Jr., eds.

111. Kline, R. and R. Kulathu. 1991. "On-Line Monitoring of Composite Prepreg Fabrication," to appear in *Proceedings ASME Winter Annual Meeting.*

112. Gruber, J., J. Smith, and R. Broukelman. 1988. "Ultrasonic Velocity C-Scans for Ceramic and Composite Material Characterization," *Mat. Eval.,* 46:90–96.

113. Belasubramanian, K. and J. Rose. 1990. "Guided Wave Potential for Damage Analysis of Composite Materials," *Review of Progress in QNDE,* 9.

114. Hsu, D. and H. Jeong. 1989. "Ultrasonic Velocity Change and Dispersion due to Porosity in Composite Laminates," *Review of Progress in QNDE,* 8:1567–1573.

115. Hsu, D. 1988. "Ultrasonic Measurements of Porosity in Woven Graphite-Polyimide Composites," *Review of Progress in QNDE,* 7:1063–1068.

116. Hsu, D. and A. Minachi. 1990. "Defect Characterization in Thick Composites by Ultrasound," *Review of Progress in QNDE*, 9:1481–1488.

117. O'Donnel, M., E. Jaynes, and J. Miller. 1981. "Kramers-Kroenig Relationships Between Ultrasonic Attenuation and Phase Velocity," *J. Acoust. Soc. Am.*, 69:696–701.